Contents

LIST OF FIGURES

This edition published in the UK in 2017 by
Icon Books Ltd, Omnibus Business Centre,
39–41 North Road, London N7 9DP
email: info@iconbooks.com
www.iconbooks.com

Originally published in 2001 by Icon Books Ltd

Sold in the UK, Europe and Asia by
Faber & Faber Ltd, Bloomsbury House,
74–77 Great Russell Street,
London WC1B 3DA or their agents

Distributed in the UK, Europe and Asia by
Grantham Book Services, Trent Road,
Grantham NG31 7XQ

Distributed in the USA by
Publishers Group West,
1700 Fourth Street, Berkeley, CA 94710

Typesetting by Born Group

Printed and bound in the UK by Clays Ltd, St Ives plc

ACKNOWLEDGEMENTS

I first learned about Copernicus and his achievement from Jerry Ravetz, a charismatic lecturer at the University of Leeds, way back when. This book, and indeed my whole career as a historian of science, owes much to his enthralling inspiration. I was given my own chance to try to make Copernicus accessible and relevant to students first of all at Wolverhampton Polytechnic and later at Hatfield Polytechnic. I am very grateful to Pat Rickwood and to Beverley Southgate who gave me those initial opportunities. Needless to say, I also owe a great deal to the authors of numerous books I have read over the years. I would especially like to thank three excellent scholars – Noel Swerdlow, Robert Westman and Curtis Wilson – whose works are too technical to be included in my suggestions for further reading but upon which I relied heavily. Jon Turney, editor of this series, and Simon Flynn, publishing manager at Icon Books, offered excellent advice and generous encouragement during the writing of the book. The end result is a great improvement on my first attempts thanks to them. Thanks also to Sandra Stafford for thoughtful, efficient and friendly copy-editing. Finally, I wish to thank my parents for their unfailing encouragement and support over the years. As a very small token of my gratitude, I dedicate this book to them.

To my mother and father, Jean and Bill Henry

• CHAPTER 1 •

YOU'D HAVE TO BE CRAZY TO SAY THE EARTH MOVES

Science or insanity?

Death and taxes are often said to be the only certainties, but there's something else. The Earth certainly doesn't seem to be moving. No matter what the scientists tell us about the way the Earth rotates on its axis, so that someone standing at the equator is moving around the centre of the Earth at a speed of over a thousand miles per hour, it all seems perfectly still. We all believe what we are told about the Earth's revolutions around the Sun, covering a distance of about 584 million miles in a year (which is another 66,000 mph or more), but we are taking it on trust – we can't feel ourselves moving.

Those of us who have read a little further, or who have watched more programmes on Discovery Channel, or who once bumped into a friendly astronomer, might also have learned that we, together with the Sun, are circling around the centre of our galaxy at an even more unimaginable speed – roughly 350,000 mph – and that the galaxy itself is whizzing through space on a trajectory resulting ultimately from the Big Bang that created our universe (see the 'Glossary' section at the end of this book). But we cannot feel any of this.

It's not just a question of how it feels either. If our senses don't tell us the Earth is moving, it is also true that, for technical purposes, it actually makes sense to assume the Earth is stationary. If you look in a textbook on navigation, for example, you'll see it assumes that the Earth is stationary, and that all the heavenly bodies are revolving around the Earth. The authors of such manuals probably know better than you or me that the Earth is perpetually performing a series of complex motions, but they also know that you don't need to know this to steer your boat by the stars. On the contrary, it makes things simpler if you assume the Earth is still and only the stars, and your boat, are moving. Therefore, it is wrong to assume that technical demands must inevitably lead us to a belief in the motion of the Earth. They don't.

So *how on Earth* did we ever come to believe in the motion of the Earth? If our senses and our common sense tell us that the Earth is not fast but steadfast, and if it is a requirement of a practically useful technical art like navigation that the Earth be considered stationary, why and how did we ever come to believe that the Earth is whizzing through space with such phenomenal speeds? If we just set aside for a minute what we have taken on trust since we were schoolchildren, the idea that the

Earth is moving just seems totally crazy. It can't *really* be moving, can it?

Yes it can, and what's more we now all believe that anyone who *denies* the motion of the Earth must be a crank or a fool. So how has it come about that it is now crazy to deny what actually seems to be a crazy idea? The short answer is that we all now recognise the intellectual power and authority of science. Even if we don't know much about the details, we know that a moving Earth is bound up with the latest astronomical and cosmological ideas, which in turn are bound up with well-established theories of modern mathematical physics. We also have a strong sense, even if we can't follow the technical demonstrations, that this same edifice of mathematical physics has led to most, if not all, of the high technology that is now such an indispensable part of our lives. It is part of our scientific worldview that the laws of nature are so all-pervasive and so interconnected with one another that to reject the claim that the Earth moves would somehow have to entail not only, say, a denial that we ever landed men on the Moon, but also that television sets work.

But, of course, there hasn't always been this kind of faith in the power and pervasiveness of scientific knowledge. Like everything else, our modern worldview has its history and its historical origins. If it is possible at all to pinpoint a single initial source from which the modern physical sciences developed and spread out, explaining more and more aspects of our world as they did so, and leading to more and more technical innovations, the most likely contender for the starting point would be the claim by Nicolaus Copernicus (1473–1543) that the Earth is in motion.

One of the main aims of this book is to explain how it was that a highly technical astronomical theory, far beyond the mathematical competence and the understanding of all but a very few people, came to have such far-reaching repercussions. Along the way we will come to see why it was Copernicus's claim (first published in the middle of the 16th century) that was to have this seismic impact, rather than one of the various earlier suggestions that the Earth might be in motion.

Nothing new under the Sun

Earlier suggestions? Yes, Copernicus's assertion that the Earth is in motion is historically the most important, but it wasn't the first. Indeed, the suggestion is almost as old as theoretical astronomy itself. Although there is abundant evidence that attempts to keep track of the heavenly bodies (and to use the knowledge of their movements for calendrical, astrological, ritualistic and in some cases navigational purposes) date back to prehistoric times, as far as we know the Ancient Greeks were the first

4

to try to understand how the heavenly bodies were moving. This marked the beginning of theoretical astronomy. At about the same time as the Greeks, the Babylonian civilisation also regarded the study of the heavens as having the utmost importance. But they seem to have concentrated exclusively on making accurate records of heavenly movements in order to discover the various repetitive cycles and thereby make accurate predictions. The Ancient Greeks, however, wanted to know what was going on in the heavens. What was it that accounted for the movements of the heavenly bodies?

Some Ancient Greek astronomers and philosophers seriously considered that the Earth might be in motion. Copernicus himself mentions some of these in the preface of his book, *On the Revolutions of the Heavenly Spheres*, which he published in 1543. Quoting from a summary of the ideas of the Greek philosophers that had been written in Ancient times, Copernicus tells his readers that Hicetas of Syracuse (5th century BC), Ecphantus the Pythagorean (*c.* 400 BC), and Heraclides of Pontus (*c.* 390–339 BC) proposed that the Earth rotated on its axis, and that Philolaus the Pythagorean (*fl. c.* 475 BC) believed that the Earth, together with the Sun and the Moon, rotated around a great fire at the centre of the world system. The Greek astronomer most associated with the idea of a moving Earth, Aristarchus of Samos (*c.* 310–230

BC), was originally mentioned by Copernicus too, although his name was inadvertently left out as the result of last-minute editing of the manuscript.

In any case, Copernicus's great innovation is hardly diminished by these scanty reports of Ancient beliefs. For one thing, the reports are merely passing comments, entirely lacking in detail. The hard work of providing the precise geometrical models required to make sense of a system in which the Sun was at the centre and the Earth in orbit still had to be undertaken by Copernicus. Besides, it's impossible to be sure, but the evidence suggests that Copernicus had already hit upon the idea that the Earth was in motion, and he then looked back to see if any of the Ancients had proposed the same idea. To us, this may seem like a strange thing to do. Why would Copernicus want to detract from his own achievement by pointing out that others had had the idea before him?

We have to remember that Copernicus lived in a different age and shared the general assumptions of his age, just as we share the general assumptions of ours. We now all believe in progress, and assume that science will lead to new discoveries and new improvements in our lives, new and previously undreamed of ways of exploiting nature for our benefit. But this attitude towards scientific knowledge developed after Copernicus's time. Nobody in his day thought about scientific progress in this way.

Wisdom was not something waiting to be discovered in the future, it was something that had once existed in the past, and needed to be recovered. This was an idea that derived essentially from religious beliefs. It was taken for granted that Adam, the first man, knew all things. This is what was meant by the comment in the book of *Genesis* (2, verses 19–20) that he named all things. If you know the name of something you know its essence, its very nature. It was believed that this knowledge began to be lost after the Fall, after the disobedience of Adam and Eve, when they were cast out of the Garden of Eden. It was not lost straight away, however, but gradually, over the succeeding generations.

This was why the period during which the Ancient Greek philosophers flourished could be regarded as a Golden Age. They were closer in time to the Fall of man and therefore knew more things, had forgotten less, than the people of later ages. If, like Copernicus, you came up with a novel idea, it was important to see if there were any hints of it in the past. If there were not, it surely indicated that your idea could not have been part of Ancient wisdom and therefore couldn't be true. So Copernicus needed to know about Ecphantus, Heraclides, Hicetas and Philolaus, and he needed to tell the readers of his book about them. This attitude to the past was taken so seriously that Copernicus's theory was often called the Pythagorean theory.

The fact that none of the claims about a moving Earth ever caught on among the majority of Greek astronomers and philosophers was not too damaging to Copernicus. It was simply a matter of suggesting that the Pythagoreans were more in tune with the original wisdom of Adam, but that most of their contemporaries had already become too ignorant to recognise it.

It wasn't only the Ancient Pythagoreans who argued for a moving Earth. There were two major statements about the possibility much closer to Copernicus's own time. The first was put forward by a French philosopher of the 14th century by the name of Nicole Oresme (*c*. 1320–82). Oresme was not a professional astronomer; what he did was simply to show that the arguments put forward by the Ancient Greek philosopher Aristotle (384–322 BC) to prove that the Earth must be stationary were by no means certain. In so doing Oresme indicated that the Earth might well be moving without us being able to notice its movement. But he makes it perfectly plain that he didn't really think it was moving. He simply wanted to show that Aristotle's arguments for the necessary stability of the Earth were not as forceful and undeniable as everyone seemed to think. Aristotle was the most dominant and influential philosopher in Oresme's day, and everyone deferred to his opinions, so Oresme's dissent from Aristotle was little more than an intellectual exercise to show that it was possible to disagree with this Ancient authority. Once again, therefore, it would be unfair to let Oresme steal Copernicus's thunder as being the first thinker to insist that the Earth is in motion.

The second medieval statement of the motion of the Earth, by Nicholas of Cusa (1401–64), was inspired in a completely different way. Nicholas, who eventually rose to the status of Cardinal in the Church, wrote an influential book called *On Learned Ignorance*, in which he insisted that the truly learned man was one who recognised the limitations of his own intellect and the extent of his ignorance. Inspired by the dire straits into which astronomy had fallen by his time (about which we'll read more later), Nicholas chose as an example the futility of any attempt to understand the real workings of the cosmos. Cosmology, in other words, is beyond the grasp of the human intellect, and the learned man will acknowledge this. During the course of this discussion, he argued that we can't even be sure of the position or fixity of the Earth. But to cast doubt on the fixity of the Earth is hardly the same as systematically setting out to reform astronomy in the most rigorous way, as Copernicus did. Nothing Nicholas of Cusa said should be allowed to undermine Copernicus's position as the first to demonstrate the motion of the Earth.

We can see, then, that although the idea that the Earth is in motion is an obviously crazy idea, it wasn't so crazy that nobody ever considered it. It was, however, crazy enough to stop the idea being taken seriously until Copernicus came along. And, of course, even he didn't have things all his own way – certainly not to begin with. Nevertheless, Copernicus's idea did prevail in the end. What we need to know is how and why Copernicus's version of this crazy idea prevailed while the earlier versions failed to have any impact whatsoever.

Astronomy rules, OK?

Now, you might imagine that it was simply a question of getting the astronomy right. After all, neither Nicole Oresme nor Nicholas of Cusa were astronomers and so were forced to talk only in terms of maybes. As for the Ancient Pythagorean astronomers, perhaps they never actually succeeded in providing an astronomy capable of predicting planetary motions; we don't actually know enough about the outcome of their efforts. So, maybe all that was really lacking was a properly worked out system of astronomy in which the Sun was at the centre and the Earth in orbit around it – in which case we can simply say that Copernicus just happened to be the first astronomer with the necessary mathematical genius to come along, and he set the world to rights. End of story.

But not so fast. We've already seen that technical astronomy in itself isn't enough to convince us that the Earth is in motion. Technical astronomy just isn't that kind of thing. What astronomy does is to provide us with a geometrical model of how the heavenly bodies are moving. The only requirements of the model are that the motions of the planets and stars that it provides should closely match (the closer the better) the recorded observations of those motions, and that the proposed movements of the heavenly bodies should be at least plausible. It happens that there is always more than one way to model the motions of the heavens. And, as present-day manuals of navigation attest, at least one way can always be based on the assumption that the Earth is stationary in the centre. Therefore, given two models to choose from, a model with a *stationary* Earth is always going to seem more plausible than one where the Earth has to be assumed to be moving.

We came to the conclusion earlier that we are willing to suspend our disbelief about the crazy notion that the Earth is in motion because we have come to trust the authority of the physical sciences. But in Copernicus's day, science did not have the same commanding authority. So what made Copernicus believe that the Earth was in motion? And why did others come to believe it too?

CHAPTER 2

WHY DID COPERNICUS SAY THE EARTH MOVES?

Heavenly orbs

We've already seen that theoretical astronomy began with the Ancient Greeks. Although there were much earlier attempts to understand what was going on in the heavens, the first major stimulus to a theoretical astronomy came from perhaps the greatest of the Ancient Greek philosophers, Plato (*c.* 428–347 BC). If every intellectual were asked to compile his or her top ten of the most brilliant and influential thinkers of all time, Plato would probably be on everyone's list. He's mostly famous for his moral and political philosophies and his influence on early Christian theology, but he's also the person who kick-started the geometrical astronomy of the Greeks.

Plato didn't do any astronomy himself, but he recognised its significance. These were times when the study of the phenomena of the night skies and the motions of the Sun had both spiritual and practical import. For Plato the concern seems to have been mostly abstract and spiritual, rather than practical. He strongly endorsed earlier Greek notions, deriving from religious impulses, that the heavens should be perfect. Among intellectual Greeks, there was a belief in a supreme deity who was assumed to have created the world. But such a deity should have created a perfect world, and if the world was perfect, it shouldn't change. Surely, if something is perfect and then it changes, it would have to become not so perfect. This kind of thinking seems to have given rise to a widely held belief among the Greeks that change was somehow distasteful. The changing world was explained, therefore, in terms of the delusion of human frailty, or as the recalcitrance of the material world that was vastly inferior to the spiritual world, or as the result of the corruption wrought on the world by petty and far-from-perfect humankind. The baleful influence of humanity could not reach up to the stars, however, and so the heavens ought to be unchanging, and therefore closer to perfection than the Earthly part of the material world.

The trouble is, they are not. The Sun moves across the sky, and it does not rise and set in the same place every day. The Moon moves in complex ways, and waxes and wanes as it does so. The stars wheel around the sky once every 24 hours, and although most of them remain in the same position on the vault of the sky as they do so, there are a few wandering stars that move in seemingly irregular ways.

The Greeks assumed, however, that the complex motions of the heavens must be to a large extent illusions generated by our point of view, and that really the heavens are unchanging. The general picture that emerged from this assumption was that the Sun, Moon, planets (the wandering stars), and the fixed stars (which moved around the sky but not in relation to one another), were all on spheres surrounding the Earth. This gave the immediate advantage that the motion of Mars, say, could be envisaged, *not* as a body moving freely through space, but as a *stationary* sphere, rotating on its axis. The movement of Mars is merely the movement of a luminous spot on the surface of a sphere, which, although it rotates on its axis, does not actually move anywhere. Instead of having to think about heavenly bodies moving through space, or changing their places, the Greeks could simply think in terms of fixed spheres rotating gracefully on their axes. The Earth was at the centre of a nest of spheres that completely surrounded and rotated around it.

Because a sphere looks exactly the same no matter from where it is viewed, the sphere could be declared to be the most beautiful and perfect geometrical shape, and the most fitting for the heavens. To reinforce the notion of the heavenly movements being as close to unchanging as possible, it was pointed out that only the motion of a sphere on its axis ends exactly where it begins. Furthermore, it was assumed that the rotation of the spheres must be perfectly uniform and unchanging in speed. As far as possible, the heavens were regarded as unchanging in their heavenly perfection.

Unfortunately, this religiously and aesthetically inspired vision of the heavenly bodies and their movements fails to fit in with some fairly obvious observations. Transient changeable phenomena like comets and meteors could be dealt with simply by assuming that they were atmospheric phenomena, taking place below the sphere of the Moon. Even today we still use the word 'meteorology' to refer to the study of weather. This is a legacy from the Ancient Greek belief that meteors and comets were so fleeting that they could not belong to the unchanging heavens and must be atmospheric. But there were other changes in the heavens that could not so easily be dismissed. The lengths of the seasons are not equal, so the Sun's movements across the sky must vary in speed; the Moon's speed also varies, as does its size. The motions of the planets are also not uniform and unchanging. Indeed, they occasionally stop then reverse their direction of travel for a while before stopping again and resuming their normal direction around the heavens. During these periods of reverse or retrograde motion they appear brighter and larger in the sky, suggesting either that they are variable or that they have moved closer to the Earth.

It's at this point that the towering figure of Plato enters the story. Plato strongly endorsed the view that the world ought not to be changeable but unchanging in its perfection. Believing the changeable material world to be largely illusory, Plato mistrusted the senses. Sight, hearing and the other senses were too easily deceived to provide us with secure information about things. Certain knowledge could only be established by the use of reason. The best means of harnessing the power of human reason, Plato believed, was by following the carefully worked-out procedures of mathematics, and in particular of geometry. Over the door of Plato's Academy, which he founded in Athens in 388 BC (and after which all subsequent Academies are named), was the legend: 'Let no one enter who is ignorant of Geometry.' Plato very much admired the achievements of Greek geometry and clearly saw its proofs, which followed inexorably from pre-established axioms and an accumulation of prior propositions established in earlier proofs, as the securest form of knowledge.

Troubled by the continual changes easily observable in the heavens, Plato threw out a challenge to his contemporaries and to subsequent generations of mathematicians and astronomers to show how the seemingly changeable motions of the heavenly bodies could be reconciled with the religious and aesthetic assumptions that they were nested spheres, centred on the Earth, moving uniformly in speed. Significantly, the phrase used to sum up this challenge was 'to save the phenomena' from their seeming imperfection. This is how it was reported by a much later follower of Plato, Simplicius (*fl.* AD 540), who wrote that 'Plato had proposed as a problem to all who consider these things seriously to investigate the uniform and regular motions by means of which planetary phenomena may be saved'. Salvation will come from geometry.

Science or art?

Believe it or not, it is possible to go a long way towards saving the phenomena in the way that Plato wished. Indeed, it's possible that Plato only threw out the challenge once he'd seen his younger colleague at the Academy, Eudoxus of Cnidus (*c.* 400–347 BC), point the way. By defining the motions of each of the heavenly bodies by the interacting motions of three or four uniformly moving spheres inside one another (and all with the same centre of rotation), Eudoxus was able to achieve the seemingly impossible. Each of the nested spheres was allocated a particular component of the motion of the body in question in such a way that the combined motions of all the spheres, each of which had its own axis of rotation so they were not all rotating in the same orientation, gave rise to a path for the planet close to that observed in the heavens. Even the retrograde motion of a planet, the period when it turns back on itself and goes in the opposite direction to its usual eastward progress across the sky, can be emulated by the use of two spheres rotating in opposite directions while being carried by the rotation of a third. A cunning arrangement of the axes and directions of rotation results in a path for the planet named by Eudoxus after a hippopede (a figure-of-eight shaped rope for hobbling horses' feet), in which the planet loops the loop.

It's easy to imagine that for a while Plato must have been genuinely excited. Like the Great Gatsby, he must have felt that his dream was so close he could hardly fail to grasp it. He evidently wasn't the only great thinker who thought the dream could be made reality. Aristotle, second only to Plato as the greatest thinker among the Ancient Greeks, and someone else who would surely figure in everyone's top ten of the most influential thinkers of all time, committed himself to a belief in a universe of nested spheres in the Eudoxan manner.

Aristotle made a few technical changes, suggesting a slightly different arrangement of spheres to those of Eudoxus, but much more importantly, he committed himself and those who chose to accept his ideas to an entirely physical, realist interpretation of the Eudoxan spheres. According to Aristotle, these spheres really were out there, carrying the planets around the sky. The physicality of the Aristotelian universe is quite explicit in Aristotle's writings and this is important because it clarifies a previously uncertain situation. We don't know for sure whether Plato believed in the real existence of the Eudoxan spheres, or whether he thought of them as figments of the mathematician's imagination, which were simply useful for showing how the movements might be understood but told us nothing about reality. Given Plato's views on the nature of what is real and what is not, it's perfectly possible that he held spheres that can only be perceived by the power of geometrical reasoning to be more real than tangible bodies. To accept the reality of such spheres, however, you have to accept the Platonic position that the material world is like a shadow-play, which only very crudely represents the real world, and that the real world itself is in fact a nonphysical, abstract realm of thought.

Not everybody can accept such a counter-intuitive notion of what is real and what is not, however. Aristotle certainly couldn't. Although for a while he was a student of Plato's at the Academy, Aristotle never adopted Plato's worldview, but soon set himself up as the champion of a completely different way of looking at the world. Aristotle's views were much more physicalist and down-to-earth. While for Plato what was real was something abstract beyond the physical world, for Aristotle what was real had to be physical and immediately present; everything else was just talk.

The Aristotelian version of the Eudoxan universe became extremely influential. Indeed, all of Aristotle's opinions became extremely influential. By a series of what were essentially historical accidents, the works of Aristotle were the only philosophical and scientific works widely known to Western Europe in the medieval period. After the decline and fall of the Roman Empire, the West entered a period usually called the Dark Ages, because it was regarded as a time when the 'lamp of learning' was 'snuffed out' (and all that kind of talk). This came to an end in the 12th century when the first universities were established and a number of Ancient writings, and more recent commentaries on them, were recovered from Byzantine Christian scholars in Eastern Europe or Islamic scholars in Sicily and Moorish Spain. For various reasons, it just so happened that Aristotle was virtually the only Greek philosopher whose writings now became available to medieval thinkers in the West. As a result, Aristotle did not even have to be mentioned by name by medieval writers. Medieval thinkers only had to refer to *the* philosopher'. The philosopher was Aristotle; there wasn't really anybody else.

According to Aristotle, the heavens were all composed of a fifth element. The four elements of earth, water, air and fire constituted all earthly bodies (in combinations of different proportions), but the heavens were made of a quintessence, or 'aether'. The Sun, Moon, planets and stars were made of aether and were carried around on spheres made of the same aether. In some versions of Aristotelianism, especially in the late Middle Ages, these aetherial spheres were thought to be solid crystalline structures, completely surrounding the Earth like the layers of an onion, but invisible due to their perfect transparency. This was simply to take Aristotle's physicalist philosophy to its logical conclusion by insisting on the solid and material nature of the heavens.

The Eudoxan universe of nested spheres, championed in their different ways by the two most influential philosophers from Ancient times, Plato and Aristotle, looked for a while like the triumph of the endeavour to explain the motions of the heavens only in terms of the uniform circular motions of perfectly spherical orbs centred on the Earth. Unfortunately, it quickly proved to be a triumph that rang hollow. Plato couldn't realise his dream after all.

Although Plato tried to play down the relevance of the actual observed motions of the heavenly bodies – he once wrote that 'what is seen in the heavens must be ignored if we truly want to have our share in astronomy' – there were simply too many obvious discrepancies between Eudoxus's theoretical account of how the heavens worked and what was actually observed for astronomers to turn a Platonic blind eye to the heavens. So attempts 'to save the phenomena' had to be continued.

The most obvious problem was the variation in size and brightness of some of the heavenly bodies (notably the Sun, Moon, Venus and Mars). If you think about it, since the Earth is wheeling around the Sun along with all the other planets, and they are all travelling at different speeds, sometimes the Earth will be on the opposite side of the central Sun from a particular planet, and sometimes the Earth and that planet will be together on the same side of the Sun. Clearly, Earth and planet will be much closer together when they are on the same side of the Sun, than when the Sun is in between them. The difference in distance has an easily visible effect on the brightness of the planets whose orbits are closest to the Earth's: Mars and Venus. What's more, the apparent sizes of the Sun and Moon varied in a way that could actually be measured. These variations are easy for us to understand, but if you are assuming the Earth is stationary and that the heavenly bodies move around in perfect circles that are centred on the Earth, it is by no means so easy. The heavenly bodies always ought to be exactly the same distance away and their appearances shouldn't vary.

The religious and aesthetic beliefs that suggested perfect spheres were the only plausible choice that the divine creator of the world system would have made, continued to hold the Greeks in thrall. It was, after all, endorsed not only by the hugely admired Plato, but also by his staunchest critic, Aristotle. And, of course, they continued to believe that the Earth was stationary.

Over the succeeding centuries, Greek astronomers tried various ways to meet Plato's challenge to save the phenomena. The outstanding figures in the story were Apollonius of Perga (*c.* 245–190 BC), who was primarily a mathematician, and Hipparchus (*c.* 190–120 BC), who was also a meticulous astronomical observer.

It was perhaps largely due to the influence of Hipparchus that the character of Greek astronomy changed at about this time. Hipparchus knew of the separate tradition of astronomy that had been pursued by the Babylonians; he is known to have relied on some of their observations. They had concentrated on keeping careful records of observations to enable them to work out what regularities there were in the various cycles of the heavenly bodies. They had not, however, tried to show how such cycles might be produced by geometrical models of the heavens.

It seems fair to say that, by contrast, the Greeks had been a bit slap-dash in their attitude to observations. Like the proverbial Irishman who would not let the truth spoil a good story, Greek astronomers before Hipparchus tended to regard the geometrical model as the important thing. If the observations didn't quite fit, that was just too bad (remember Plato's insistence that 'what is seen in the heavens must be ignored if we truly want to have our share in astronomy').

As the Greeks became increasingly aware of the rich fund of observational data collected by the Babylonians, they must have begun to see it as the way forward. Hipparchus was evidently one of the first – and certainly one of the most gifted – Greek astronomers to exploit the rich data of the Babylonians in the service of geometrical astronomy. Such major changes in approach do not come about as the result of the mere whim of one man, however, nor even as the result of a collective whim by astronomers. It seems safe to assume that part of the reason for newly acknowledging the importance of accurate observation was due to the realisation that the Platonic challenge had still not been met, and that maybe (contrary to the expectations of Plato himself) observational data would actually help. But there was also something else.

Greek civilisation was becoming increasingly aware of the practical importance of astronomy.

We know that the Babylonian interest in being able to accurately predict repetitive cycles in the heavens was driven by calendrical and agricultural concerns, and by a concern with astrology. If the heavens can tell you the most propitious time to plant and to harvest your crops, why may they not also indicate the best time to marry or to undertake an unavoidable but perilous journey? Among the Greeks, the underlying concern for the practical usefulness of astronomy in everyday life becomes very evident in the work of the astronomer who marks the culmination of the Ancient Greek tradition of astronomy, Claudius Ptolemy (c. AD 100–170). What Ptolemy did was to draw up a massive compendium of the best mathematical techniques for calculating the motions of the heavenly bodies. In doing so he digested the work of his predecessors and showed how their theories could be put to use for calculating solar, lunar and planetary positions at any time in the past or the future. This is just what is required for practical considerations. Take the case of astrology. If you wish to draw up the horoscope of a 40-year-old man, you will need to be able to calculate whereabouts in the sky the heavenly bodies were at the moment of his birth (assuming you know this information). Similarly, if the successor to a kingdom wishes to know the best date to hold his coronation, an astronomer will be required to work out the forthcoming positions of the heavenly bodies. As well as being able to work out planetary positions, of course, you will also need to know what these positions mean. Is it good for Mars to be dropping below the horizon on the date in question, or is it bad? Is it good for the Sun to be crossing the constellation of Aries, or not? Ptolemy took care of this sort of thing in another massive compendium. The fact that his great *Mathematical Syntaxis* (or, as it has been known to Western civilisation ever since the Middle Ages, the *Almagest)* is a companion piece to an encyclopaedia of astrology, the *Tetrabiblos*, shows that Ptolemy's principal concern was with the practicalities of astronomy rather than with the niceties of the geometry of the heavens.

One way of seeing the shift in Greek astronomy from the Platonic ideal to Ptolemaic pragmatism is to think of it as a shift from science to art. I mean 'art' here in the original Greek sense, which is still in use in expressions like 'the art of cookery' or 'the art of motorcycle maintenance', and *not* as in the expression 'You can tell this painting is great art because it just fetched a million pounds at auction'. The *art* of astronomy is concerned with the skilful and accurate calculation of planetary positions for practical purposes; the *science* of astronomy is concerned with understanding how the heavens work and discovering what is actually going on in the sky above us. The Greeks had another word for this: cosmology. The Greek word *cosmos* means 'order' as well as 'world' or 'universe', and usually implied that the universe was a unified and harmonious whole. Cosmology was the study of the order and harmonious arrangement of the world.

Ptolemy and the decline of cosmology

In compiling his great compendium of astronomical models and geometrical techniques for establishing planetary movements, Ptolemy drew more frequently on comparatively recent developments. Astronomy, after the merging of the Greek geometrical tradition with Babylonian quantitative observation, proved most useful to him. Accordingly, he moved further and further away from the picture of the world as a perfectly neat nest of concentric spheres, as described by Eudoxus and made physical by Aristotle.

Drawing on ideas first developed by Apollonius, for example, Ptolemy used what are known as eccentrics and epicycles to account for observed variations in speed of the planets, their variations in appearance (size and brightness) and their retrograde (backwards) motions. The eccentric was simply a presumed displacement of the Earth from the centre of the planet's sphere (see Figure 1). If the Earth is not at the centre of the sphere's rotation it means that some parts of the sphere must be closer to the Earth than other parts. This will result in a changing appearance of the planet as it moves nearer to, or further from, the Earth. It also means that the

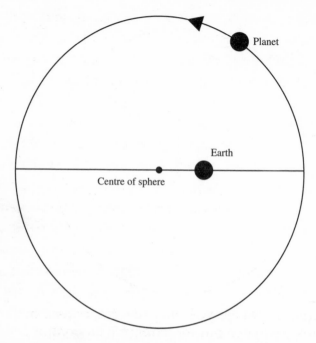

Figure 1. The eccentric, as used by Ptolemy to partially explain seeming variations in the speeds and appearances of the planets.

motion of the planet will seem to speed up and slow down. Even though the planet's speed around the centre of its sphere is constant, the further it is from Earth, the slower it will seem to be moving (if you have trouble envisaging this, compare the sensation of a car travelling at 30 mph just past your nose with one that you observe from a hill-top a mile away).

The epicycle is slightly more complicated. The planet is assumed to move on a circle (the epicycle),

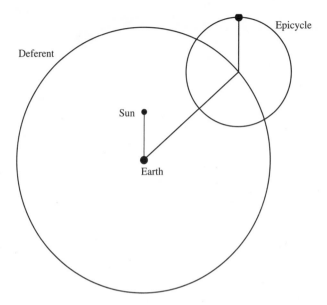

Figure 2. The epicycle on a deferent, as used by
Ptolemy to account for retrograde motions and
other seeming irregularities of planetary motions.
Note that, for the outer planets, the line joining
the Earth to the Sun was always parallel to the line
joining the planet to the centre of its epicycle. This
means that the planet always turns on its epicycle
at the same rate as the Sun turns around the Earth.

which is itself moving around a larger circle (the
deferent) (see Figure 2). By setting the required
values of the speed of the planet on the epicycle
and of the epicycle around the deferent it is possible
to make the planet loop the loop. The period when
the planet is on the part of the epicycle inside the
deferent will be a period of retrograde motion. It

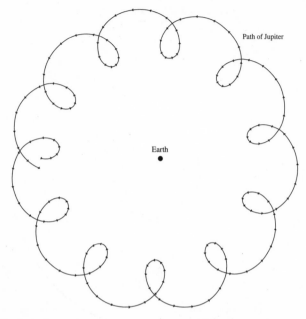

Earth
●

Path of Jupiter

Figure 3. This shows how the combination of motions on epicycle and deferent result in a path that accounts for the retrograde motions of the planets – in this case – based upon observations between 1708 and 1720 – Jupiter. It can be seen at the left that the planet, moving anti-clockwise, does not quite get back to its starting point after 12 years, according to the Ptolemaic scheme.

can also be seen to be the period when the planet is closest to the Earth (see Figure 3). This is exactly what the observations demand. Retrograde motion takes place when the planet appears to be at its closest. Indeed, we now know that the retrograde motion of a planet is an optical illusion caused by

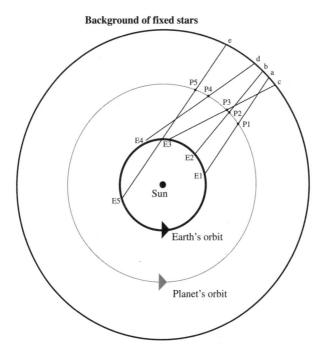

Background of fixed stars

Figure 4. Copernicus's explanation for retrograde motion. The diagram shows five successive positions of the Earth (El, E2 and so on), and five successive positions of the planet (Pl, P2 and so on). Although the planet only moves in one direction, the line of sight from the Earth (overtaking it on the inside) makes it look as though the planet goes from a to b, then back to c and on again to d and e.

the Earth overtaking the planet as it moves around its own orbit, which it does at the point when it is closest to the planet (see Figure 4).

These cunning innovations were generally held to be in keeping with religious and aesthetic demands, in so far as the motions of the heavenly bodies were still assumed to be perfectly circular, and perfectly uniform and unchanging in speed (the speed of the planet might seem to vary when observed from Earth but it was uniform about the centre of rotation). But already the picture presented by eccentrics and epicycles is by no means as neat and immediately plausible as the Eudoxan/Aristotelian universe. There's no denying the ingenuity of the epicycle, particularly when its fit with observations is considered. But could a planet really move around an empty point in space, which is itself moving? Quite a few earlier thinkers were suspicious of this idea.

What was far worse, however, was another innovation that Ptolemy himself devised. The perceived variations in speed of a planet could not always be accounted for simply by supposing a centre of rotation eccentric from the Earth. So, Ptolemy introduced another significant geometrical point into the picture; he called this the equant point. What we have here is a point in space on the opposite side of the centre of rotation of the planetary sphere from Earth's position. What is this point for? It marks the position in space from which the planet would appear to be moving with uniform circular motion.

Let's look at that again. The speed of rotation of the planet would look uniform and perfectly regular if it could be observed from the equant point (which is, in fact, an empty point in space).

It follows from this, as sure as night follows day, that the planet *cannot* be moving with uniform circular motion around the centre of its own sphere or deferent. Its speed around its own centre must be varying, getting quicker as the planet moves further away from the equant point and becoming slower as it approaches it. Accordingly, the variation in speed will seem all the greater when viewed from the Earth (again being faster nearer the Earth and slower further away – see Figure 5).

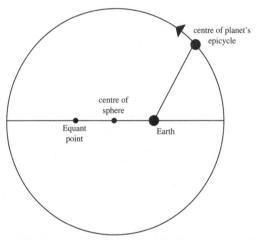

Figure 5. Ptolemy proposed a point in space, called the equant point, and supposed that the motion of the planet, or the centre of its epicycle, would look uniform in rotational speed when seen from that point. This works very well in explaining the actual observed movements of the planets, but it implies that the planet's sphere must be alternately slowing down and speeding up as it rotates around its centre. This was considered to be physically impossible.

Ingenious though this was, for most astrono-mers and natural philosophers from Ptolemy's time to the Renaissance it was a highly regret-table departure from the Ancient stricture to save the phenomena in terms of uniform rotations in perfect circles. Motion that was uniform around the equant point was not uniform motion around a circle. Although the equant worked wonderfully well, fitting the observed motions better than any previously tried geometrical models, there was still a general feeling that Ptolemy was a cheat and a cad. If the Ancient Greeks had played the game, it would have been widely asserted that Ptolemy's astronomy just wasn't cricket.

But some things never change, and then, as now, what can readily be put to use (especially if a personal profit is to be made) tends to win out over rarefied theory. So Ptolemy's astronomy, thanks to its success in calculating heavenly motions, carried the day and came to be recog-nised (especially by professional astronomers and astrologers) as the acme of Ancient astronomy. The fact that Ptolemy had cheated and deviated from the uniform circular motion laid down as a golden rule by earlier thinkers was for the most part quietly overlooked. We can see this, therefore, as the triumph of the pragmatic *art* of astronomy over astronomical or cosmological theory.

How to avoid a crisis: the medieval stand-off

Now, to get back to Copernicus. With the fall of the Roman Empire in the 6th century AD, Western Europe entered the Dark Ages and the wisdom of the Ancient Greeks was lost. But things began to pick up in the 12th century when a number of important Greek works in philosophy, medicine and astronomy were recovered from Islamic or Eastern European sources (where the legacy of the Ancient Greeks had to some extent been continuously preserved). The recovery of Ancient wisdom caused great excitement among those who had a thirst for knowledge and learning, and the study of these materials was incorporated into the curricula of the first universities, which were established at this time. It was from this time on that the Ancient Greek philosopher Aristotle was referred to as *'the* philosopher'. It just so happened that the only significant body of philosophical writings available to the West at this time were his writings. Accordingly, they were venerated, pored over, memorised and generally accepted as true.

Ptolemy's *Almagest* was also recovered at this time from the Arabs (indeed, its common title derives from its Arabic title, not from the original Greek), but it was mathematically far too difficult and at first its main principles had to be presented in a few simplified textbooks for use in the universities and by astronomers. Only very gradually over the succeeding centuries did mathematical knowledge improve sufficiently for the *Almagest* to be studied and used in its own right. Even then it was too much for all but the very best mathematicians and a shortened version of it, the *Epitome* (1496), was produced by the two leading astronomers of their day, Georg Peurbach (1423–61) and Johannes Müller (1436–76).

This brings us to the eve of the historical moment when Copernicus takes up astronomy. So, let's take stock of the situation at this time.

Aristotle's philosophical system is completely dominant. All university philosophy curricula were based on his works for the simple reason that there was no alternative. Aristotle's works were encyclopaedic in their scope and anyone trained at a university anywhere in Europe would be steeped in the medieval version of his philosophy. This meant, therefore, that the Aristotelian world picture was what everyone believed. The Earth was in the centre, surrounded by a series of neatly nested heavenly spheres, like the layers of a perfectly spherical onion, and these spheres were moving around their axes with perfectly uniform, unchanging, motions. Remember, this was essentially the picture of the universe developed by Eudoxus in response to the challenge thrown out by Plato to 'save the phenomena'. We've already seen that this neat picture can't fit the facts, but Aristotle did not know that. So, all educated Europeans believed in this simple picture of the universe (see Figure 6). The general assumption was that this is the way the universe really is.

Figure 6. A 16th-century rendering of the Aristotelian universe. Around the central Earth it depicts the sphere of air and the sphere of fire (the belt of flames just below the Moon), showing the Aristotelian assumptions about the natural places of each of the elements. The words around the outside read 'The ruling heaven, dwelling place of God and all the Elect', showing how this picture of the universe dovetailed with religious beliefs.

So, how did this fit with Ptolemy's astronomy? Essentially, it didn't. There had been attempts throughout the Middle Ages to complicate the Aristotelian picture in order to accommodate epicycles and eccentrics. The Eudoxan spheres were replaced by spherical shells of varying thicknesses, each of which contained embedded within them a smaller sphere that was the physical embodiment of the planet's epicycle. But this was generally seen as a compromise, a fudge, and rarely satisfied either Aristotelian philosophers or professional astronomers.

A major part of the problem for reconciling Aristotle and Ptolemy was the way the *Almagest* was written. Remember that it was a companion piece to Ptolemy's *Tetrabiblos*, his compendium of astrological lore. Ptolemy's aim was to provide professional astrologers and astronomers with the means of calculating the positions and movements of the heavenly bodies. In order to do this he took Sun, Moon, and each of the planets separately, demonstrating the way to use the geometrical model for each one in order to build up a picture of where each body was in the sky at the precise moment of the horoscope being drawn up. The point about this was that there was absolutely no attempt to consider the system as a whole. The *Almagest* dealt with each of the heavenly bodies in turn and then its job was done. Ptolemy did not go on to show how they all fitted together, nor how they could be reconciled to the Aristotelian world picture. (In fact, Ptolemy did try to do that in another work, the *Planetary Hypotheses*, but the complete version of this was only discovered very recently, in 1967.) As far as readers of the *Almagest* were concerned, Ptolemy simply provided geometrical methods for calculating heavenly movements; there was never any discussion of physical reality, of the real structure of the heavens.

More importantly, Ptolemy's failure in the *Almagest* to show how his theories fitted in with those of *the* philosopher, Aristotle, was not just an oversight that could be easily supplied by any of the book's readers. On the contrary, the technicalities seemed to suggest that Ptolemy's astronomical models *could not be used to establish what the heavens were really like*. In particular, Ptolemy's scheme could not even establish the order of the planets. Providing that the mathematically necessary proportions between a planet's deferent and its epicycle were maintained, any planet could be placed immediately outside the Moon's sphere, or immediately below the outermost sphere of the fixed stars or anywhere in between. This is what Copernicus's friend Georg Joachim Rheticus (1514–76) meant when he wrote in his *First Account* (1540):

What dispute, what strife there has been until now over the position of the spheres of Venus and

Mercury, and their relation to the Sun. But the case is still before the judge. Is there anyone who does not see that it is very difficult and even impossible ever to settle this question while the common hypotheses are accepted? For what would prevent anyone from locating even Saturn below the Sun, provided that at the same time he preserved the mutual proportions of the spheres and epicycle, since in these hypotheses there has not yet been established the common measure of the spheres of the planets, whereby each sphere may be geometrically confined to its place?

This powerfully reinforced the impression that the *Almagest* consisted merely of a set of mathematical means of analysing heavenly movements without offering any understanding of what the cosmos was really like. As Copernicus himself put it in the preface to his life's work *On the Revolutions of the Heavenly Spheres*, Ptolemaic astronomers could not even solve 'the principal consideration',

... that is, the structure of the universe and the true symmetry of its parts. On the contrary, their experience was just like someone taking from various places hands, feet, a head, and other pieces, very well depicted, it may be, but not for the representation of a single person; since these fragments would not belong to one another at all,

a monster rather than a man would be put together
from them.

Ptolemy, in Copernicus's account, seems to have prefigured Dr Frankenstein in the building of monsters. What Ptolemy did was merely a set of exercises in the art of astronomy, which could make no contribution to cosmology.

This had been recognised even in the 12th century by one of the greatest of the Islamic philosophers, Ibn Rushd (1126–98), known in Latin as Averroes, whose massive commentaries on the works of Aristotle earned him the sobriquet *'the* commentator'. Right from the outset of the revival of learning in Western Europe, therefore, intellectuals were likely to be influenced by Averroes's *Commentary on Aristotle's Book on the Heavens (c.* 1180). In this he writes that:

[t]he astronomical sciences of our days offer
nothing from which one can derive an existing
reality. The model that has been developed in the
times in which we live accords with the computa-
tions not with existence.

There were a number of other technical problems that added to this lamentable view of Ptolemy's astronomy. The most glaring of these was the uncomfortable awareness of astronomers that

Ptolemy's use of the equant point was a mathematically inspired deviation from what the Ancient Greeks had believed was physically possible in the heavens (namely uniform circular motion). Another obvious problem was Ptolemy's own attitude to his theory of the Moon's movement. This geometrical model implied that the Moon's size ought sometimes to appear double what it is at other times, which it never does. Ptolemy blithely ignored this. For him the important thing was that his model provided the right movements of the Moon across the sky. But the fact that it did so in a way that was incompatible with the physical appearance of the Moon proved hard for those in the know to ignore. It is pretty clear, then, that there was something of a clash between Aristotelian cosmology and Ptolemaic astronomy. But this clash, evident since the 12th century, had not resulted in a crisis that undermined either Aristotle or Ptolemy. Both continued to be taught alongside each other in the universities and their incompatibilities were simply condoned. There was, therefore, a stand-off.

Immediately, we have to ask how this stand-off could continue for about 500 years without either side trying to break the deadlock. History does not usually stand still, so when a situation doesn't change it calls for an explanation just as much as when things do. There are two main reasons for the continuity of this stand-off, both intertwined with one another.

On the one hand, it was maintained thanks to an important principle of Aristotle's teaching. Aristotle insisted on separating physics from astronomy as essentially different kinds of intellectual pursuits. Physics, he insisted, was a legitimate part of philosophy, but astronomy was not. Philosophy was characterised by the fact that it provided explanations for things in terms of reasons and causes. The distinction between physics and astronomy could easily be maintained, therefore, because physics gave explanations in terms of real causes but astronomy gave explanations in terms of geometry. It is one thing to explain the motion of a heavenly sphere by attributing an intelligence to it, as the Aristotelians did, so that the intelligence could move the sphere in the same way that we can move our own bodies, by willing it to happen. It is quite another to say that an epicycle moves around a deferent in order to make a planet loop the loop (as it performs retrograde motion). The former explains why the planet actually moves. However, the latter only shows how it can perform the loop; it leaves the physical reason for its movement, the actual cause of motion, completely out of the reckoning. For Aristotle, this meant that astronomy was inferior to natural philosophy because it was essentially descriptive, rather than explanatory. This tended to reinforce the distinction between the *science* of the world system, or cosmology, and the merely practical *art* of astronomy.

This purely intellectual distinction was to have institutional and professional ramifications. The professional natural philosophers – that is to say, the professors of philosophy in the universities – were able mostly to ignore the mathematical technicalities of astronomy, which seemed to suggest that the Aristotelian world picture must be wrong. At a pinch, they could refer to the compromise picture, in which epicycles were pictured as being embedded in eccentric spherical shells rolling around the Earth. The aim here was clearly to forestall awkward questions from bright students who might notice that astronomy and cosmology didn't quite match up. If the philosophy professors did occasionally acknowledge the difficulties presented by astronomy, they only had to repeat the age-old challenge that it was up to the mathematicians to show how the observations could be reconciled with what everyone assumed to be the physical reality of heavenly spheres, centred upon the Earth, and moving with uniform rotations around it. Since even Ptolemy had failed to do this, the professors could throw out this challenge without any fears that it would result in a duel that they might lose.

Professional astronomers, on the other hand, could carry on doing what they did best – namely calculating heavenly movements or showing others how to make the necessary calculations. If this involved the use of what was supposed to be a physically implausible model like the equant point, or turning a blind eye to the physical appearance of the planet, so be it.

If astronomy could not provide physical explanations of heavenly phenomena, as the great Aristotle had said, then it hardly mattered if astronomers sometimes proceeded as though physical considerations were irrelevant. Within the university system, after all, what astronomers taught was primarily intended to enable medical students to be able to cast horoscopes. One of the first things any doctor would do when dealing with a new patient would be to use astrology to help him to reach a diagnosis and a prognosis of how the illness might proceed. Astronomy and astrology were very important aspects of medical training, therefore. The concerns here were entirely pragmatic. Astrology was used in medicine because it worked – or so most people thought – and that suggested that the astronomers were doing something right. Patients usually got better after medical treatment, and if that treatment was partly based on astrology then astrology couldn't be nonsense (in fact, these patients probably would have got better anyway without the benefit of medicine or astrology, but that wasn't realised at the time). So, if what the astronomers taught didn't quite match Aristotelian natural philosophy it was easy to overlook it.

The institutional separation between philosophers and astronomers was also clearly marked in their salaries. Professors of natural philosophy routinely received much higher salaries than professors of mathematics. This is strikingly apparent, even in the case of a mathematician as famous as Galileo Galilei (1564–1642). His salary as professor of mathematics at the University of Padua was increased from 520 florins a year to the exceptionally high sum of 1,000 a year as a result of his famous discoveries with the telescope. However, Galileo's colleague, Cesare Cremonini (1550–1631), a professor of natural philosophy, was granted 2,000 florins a year. There can be no question that mathematicians were regarded as inferior to philosophers.

Factors like these made it possible for the stand-off between philosophers and astronomers to last for so long. Throughout the Middle Ages and up to Copernicus's day, mathematicians had made various attempts to improve the accuracy of Ptolemaic astronomy, but there were no significant attempts to resolve the tensions between the Aristotelian world picture and Ptolemaic astronomy. Nothing much changed, therefore, until Copernicus came along.

How to create a crisis: Nicolaus Copernicus, cosmologist
It seems safe to assume that Copernicus began his work in astronomy with no greater ambition than

to add to the repertoire of refinements to Ptolemy's original, with a view to making astronomy more accurate or perhaps to simplify some of the calculations for the benefit of less skilled practitioners. But Copernicus was to burst through the confines of medieval astronomy and was to become a cosmologist in the Ancient mould, like a new Eudoxus seeking to save the phenomena for his revered Plato.

It is impossible to know for sure exactly how Copernicus came up with his new picture of the world system, with the Earth in motion round a central Sun, but he tells us himself that it was his attempts to do away with Ptolemy's use of the equant point, which 'gave us the occasion to consider the mobility of the Earth'. As we've seen, the equant was the major scandal that blighted Ptolemy's otherwise admirable achievement; it was regarded as an unforgivable deviation from the stricture on astronomers to explain heavenly motions in terms of perfect and unchanging circular motions. According to Copernicus's friend, Georg Rheticus, it was 'a relation that nature abhors'.

Copernicus first succeeded in accounting for the movements of the outer planets (Mars, Jupiter and Saturn) without recourse to the equant by using an epicycle on an epicycle. These two small epicycles accounted for the observed variations in speed of the planets (in the same way that the equant did), but this was to add two small epicycles to the large one that Ptolemy was already using. Furthermore, Copernicus's extra epicycles introduced an irregularity into the movement of the planet with regard to the centre of the deferent. By trying to get rid of one irregular kind of movement, implied by the use of the equant, Copernicus had only succeeded in introducing another. On examining this new irregularity, however, it seems likely that Copernicus noticed that he could avoid it by transferring the motion of the planet on its large epicycle to the Earth itself.

It may have been that he came to this idea only after first transferring some planetary movements to the Sun. Earlier astronomers had already suggested that Mercury and Venus might be going around the Sun, since their deferents turned at the same rate as the Sun. The outer planets, by contrast, had deferents with different rotational periods, but they all moved around their epicycles in one year (refer again to Figure 2).

To simplify the system, Copernicus could make the outer planets more like the inner planets by making the Sun's sphere act as their deferents too, and having their epicycles move at the necessary different rates. This meant that all planets turned about the Sun while the Sun turned about the Earth. If Copernicus did follow this path he may well have been sufficiently excited by such simplifications to notice the final simplification of making the Sun stationary and putting the Earth into orbit between Venus and Mars.

We can't be sure that this is how Copernicus did it, so in a sense these technical details aren't important. The important thing to note is simply that Copernicus was the first to realise that a set-up in which the Earth moved with the other planets around a stationary Sun was *geometrically equivalent* to the Ptolemaic system, which is to say that it accounted for all the observations.

Although we can again only speculate, it seems reasonable to assume that Copernicus was such a gifted astronomer that where others might have immediately jettisoned this crazy idea he was able to see a number of other remarkable advantages arising from the hypothesis. This would have included benefits arising from supposing the Earth was moving in other ways. If it's moving around the Sun anyway, why not assume that it's also spinning on its axis? This would then, at a stroke, dispose of the need to assume that while the planets make their way around the sky at greatly different speeds in an easterly direction, they are simultaneously moving in a westerly direction, together with the sphere of the fixed stars, once every 24 hours. This westerly motion could now be seen as merely an illusion caused by a daily eastward rotation of the Earth on its axis. Here again, the motion of the

Earth immediately makes the rest of the system much simpler and more harmonious.

As well as wheeling around us every 24 hours, the whole sky was also known to move very slowly in an easterly direction. This barely perceptible motion, first noticed by the Ancient Greek astronomer Hipparchus, was interpreted as a westward movement of the equinoxes, the two points where the Sun crosses the celestial equator each year.

This so-called precession of the equinoxes could again be much more easily accounted for by Copernicus. In his new system, precession was explained by assuming that the axis of the Earth's rotation didn't always remain in the same orientation but gradually moved around. The spinning Earth has to be imagined like a spinning top as its motion dies down – the top does not spin bolt upright but leans over and the direction of its lean continually moves around in a circle. Precession of the equinoxes can be explained by assuming the Earth moves like the top, but much, much slower (changing its direction around the compass by 1 degree every 73 years).

Although these new solutions to old problems were immediately impressive, there can be little doubt that for Copernicus the most important aspect of his new system stemmed from the Earth's revolution round the Sun. It was this motion that had the most important cosmological implications.

The annual revolution of the Earth meant that the order of the planets, arbitrary in the Ptolemaic system, could be established easily and unequivocally. In short, Copernicus was able to present his new astronomy not just as a disconnected way of calculating the movements of each of the heavenly spheres, but as a cosmology, a unified account of the way the world system really is arranged. As Copernicus himself proudly stated it at the beginning of his book *On the Revolutions of the Heavenly Spheres*:

> *Having thus assumed the motions which I ascribe to the Earth … by long and intense study I finally found that if the motions of the other planets are correlated with the revolving of the Earth, and are computed for the revolution of each planet, not only do their phenomena follow therefrom, but also the order and size of all the planets and spheres, and heaven itself is so linked together that in no portion of it can anything be shifted without disrupting the remaining parts and the universe as a whole.*

With a dazzling stroke, therefore, Copernicus had brought to an end the begrudging, centuries-long stand-off between astronomers and cosmologists. He had done so by showing that astronomy could be used, in spite of what Aristotle said, to establish a cosmology, a picture of the way the universe really is.

Unfortunately, the cosmology that Copernicus's astronomy established was hardly what was expected. The assumption had always been that, if the stand-off was to be reconciled, the result would be an affirmation of the Aristotelian universe or something so close to it that it made no difference. If Copernicus was right, it was going to have to make a very big difference indeed. For one thing, the whole of Aristotelian physics would have to be replaced.

In view of what we said earlier about the inferiority of astronomy to natural philosophy in the hierarchical structures of the universities, our expectation might be that Copernicus would give way, and defer to the insistence of the Aristotelian philosophers in the universities who could use the principles of physics to prove that the motion of the Earth is absolutely impossible. The fact that Copernicus did not defer to this privileged group points finally to the main reason why he was able to insist that, in spite of virtually everything indicating the contrary, the Earth must be in motion.

Quite simply, Copernicus refused to acknowledge the superiority of natural philosophers over mathematicians. In the preface to his epoch-making book he squarely lays the blame for the failure of cosmology and astronomy to present a unified picture of the cosmos on the philosophers:

I began to be annoyed that the movements of the world machine, created for our sake by the best and most systematic Artisan of all [he means God], were not understood with greater certainty by the philosophers, who otherwise examined so precisely the most insignificant trifles of this world.

It was obvious to Copernicus that the reason for this failure was not only because they had allowed themselves to be misled by Aristotle, but also because they had been unable, and unwilling, to pursue the correct road to discovery of the reality of the cosmos – namely, geometrical astronomy. Both parties in the medieval stand-off had been in the wrong. To get beyond the stand-off it was not sufficient to revise astronomy to fit in with Aristotle's world picture; both the old cosmology and the old astronomy had to be rejected. This had been made clear, at a stroke, by Copernicus's new astronomy.

For Copernicus, then, it was not philosophy that pointed to the truth about the nature of the heavens but astronomy. Where Aristotle had insisted that only philosophy could provide reasons and causes for the heavenly motions, Copernicus believed otherwise. For most of his contemporaries, steeped as they all were in the Aristotelian view, mathematics was held to be irrelevant to determining the truth about the way things are.

Copernicus's position, therefore, must have seemed completely incomprehensible to almost everyone. For him, however, it was perfectly simple: *the Earth was in motion because the mathematics demanded that it must be*.

Today we are entirely used to the idea that mathematics describes and defines the nature of the world around us. But our way of thinking has its origins in the work of Nicolaus Copernicus. Before his time, the notion that the world was in some fundamental way mathematical, which had been held by the Ancient Pythagoreans as well as Plato and his followers, had been lost. It had been replaced by an entirely qualitative philosophy that speculated on the nature of things and tended to offer explanations of things based on assumptions that derived from what we call common-sense. For these thinkers, mathematics was just far too abstract to be relevant to a proper understanding of brute reality. Copernicus, however, can be seen as a thinker who returned to the Pythagorean and Platonic way of thought, in which number and geometry provided the keys to understanding the universe. Twentieth-century physics can be seen as the apotheosis of this view. When asked what he would have done had a particular experiment not confirmed his general theory of relativity, Albert Einstein (1879–1955) said: 'Then I would have been sorry for the dear Lord – the theory *is* correct.' Similarly, Paul Dirac (1902–84) claimed that he knew his prediction of the existence of antimatter must be true because the mathematics was so beautiful. Antimatter had to exist because the mathematics demanded it. Twentieth-century physical scientists who rely equally on mathematical predictions and proofs in their physics, are the heirs of Copernicus. Although Einstein once referred back to the Ancients in support of his view – 'I hold it true that pure thought can grasp reality, as the Ancients dreamed' – his was really an attitude that has had a continuous history in Western science only since the revolutionary insistence of Copernicus that geometry can grasp the reality that the Earth goes round the Sun.

• CHAPTER 3 •

WHO WAS COPERNICUS?

Why Copernicus?

So far, we have only a partial answer to our question. We wanted to know why Copernicus came up with the crazy idea that the Earth is moving. We've seen the immediate reason for it: because he believed that the geometrical astronomy that clearly pointed to a coherent cosmology must be physically true. And it must be physically true for the very reason that it provides a coherent cosmology. We spent a long time getting to this answer in order to make it clear that although such an attitude might now look fairly commonplace in so far as it's the kind of argument that modern physicists would routinely use, in Copernicus's day it was an extremely radical and revolutionary move.

What has not been explained, however, is why it was Copernicus who made this revolutionary move, and not some other astronomer. It won't do to say it was because Copernicus was a genius. This is no kind of explanation at all. We could say that all innovators do what they do because of their genius, but we might just as well say that they do it because they do it. So, we need to look at the historical context to see if we can find reasons that might have led Copernicus to develop his revolutionary attitude towards mathematical knowledge, and that therefore enabled him to declare the Earth to be in motion purely on the grounds that the geometry of a universe with a moving Earth was more coherent than one in which the Earth was stationary.

Life and times

Copernicus was born in Torun, in the Kingdom of Poland, on 19 February 1473. His father, also called Nicolaus, was a successful merchant from Cracow who married Barbara Watzenrode, daughter of a wealthy and influential family in Torun. Copernicus was ten years old when his father died and from then on he, his mother and three siblings were in the care of his uncle, Lucas Watzenrode.

In 1491 Copernicus enrolled at the famous Jagiellonian University in Cracow. Astronomy would have been an important part of his studies. It was one of the seven liberal arts that formed the basis of the introductory curriculum at all universities, but it seems to have played a larger role at Cracow than at other universities. In 1492 the author of what was effectively an early travelogue said of the Jagiellonian University that 'the science of astronomy stands highest there, and there is no more renowned school'.

Copernicus left the university when his uncle, now Bishop of Warmia, nominated him for the post of canon at the Cathedral of Frombork (Frauenburg) on the coast of the Baltic Sea. In spite of his uncle's influence, Copernicus's application did not succeed, so he decided instead to enroll at Bologna University in 1496 to study law. Copernicus was evidently a keen and hard-working student. Although officially at Bologna to study for the vocational degree in law, he threw himself into subjects of less immediately practical value. He learned classical Greek, for example, then used it to study the writings of Plato and other Ancient Greek works, which had only recently been rediscovered by Renaissance scholars. Copernicus was thoroughly excited by the Renaissance of Ancient learning. After centuries when the only Ancient philosopher whose works were available was Aristotle, the works of other Ancient writers were suddenly recovered. The scholars who recovered these Ancient writings gave the period its name – they saw themselves as living in a period of rebirth, renaissance, of Ancient wisdom.

After the accidental discovery of some Ancient writings, which had been preserved over the centuries in monastery libraries, scholars began to make systematic searches in monasteries throughout Europe. It quickly became apparent that, in many if not all of them, the monks' veneration for the written word was so pronounced that, even if nobody was much interested in reading the item in question, whenever a scroll or manuscript in their library was noticed to be deteriorating, they would make a fresh copy of it. In this way, documents acquired in the early centuries of Christianity had been continuously preserved.

The territories of the Eastern Orthodox Church, centred upon Byzantium or Constantinople, proved a rich source of Ancient Greek writings and these became more readily available in the West when Greek scholars fled from Byzantium as the Islamic Turks threatened the city before finally seizing it in 1453. These were exciting times for Western intellectuals. Philosophers whose names were only known because Aristotle rejected their views in his own writings, suddenly became available in their own words. What's more, thanks also to that other great Renaissance innovation, the printing press, these newly discovered works could be compara- tively easily disseminated to scholars throughout Europe.

The impact on intellectuals and ways of thinking throughout Europe was incalculable. Where Aristotle had once been *the* philosopher, it was now becoming increasingly apparent that he was just *a* philosopher. Consequently, there was a new recognition that knowledge should not be based on Ancient authority. Although there was still a strong assumption that the ancients must have known more than the moderns, since they were closer in time to Adam who had once known all things, the range of alternative philosophies to be found among the newly discovered writings of the Ancient Greeks indicated that Ancient authority did not speak with one voice. These alternative views might suggest that Aristotle was wrong, but who was to say which of the alternatives was right? Perhaps the only way to be sure of the truth was to try to discover it for yourself.

Northern Italy, especially Florence and Venice, was at the forefront of this process of recovery of Ancient wisdom and Copernicus was well placed during his time in Bologna to turn himself into a thoroughly Renaissance man. This certainly meant familiarising himself with the leading alternatives to Aristotle among the latest discoveries, but it also meant developing a new scepticism towards authority and a restlessness in the search for certain knowledge.

Copernicus continued to study astronomy in Bologna, assisting one of the most famous astronomers of the day, Domenico Maria de Novara (1454–1504), in his nightly observations. By 1500, when Copernicus stayed for a time in Rome, he was sufficiently expert to be able to give a lecture course on astronomy, which attracted large audiences. If the report of Copernicus's friend, Rheticus, is to be believed, it must have been an advanced course because it was attended by 'a throng of great men and experts in this branch of knowledge'. These must have been happy times for Copernicus since he seemed very reluctant to leave Italy. His uncle had finally succeeded in having him appointed as Canon of Frombork some time before, and in 1501 he had to return to Poland to confirm his acceptance of the post. Straight away, however, he applied for leave of absence to study medicine. He returned immediately to Italy – this time to Padua, probably the best medical school in the world at that time. But it wasn't long before the crunch came, and in 1503 he had to return home to justify the support he had been given to pursue his studies.

Before returning home he paid a brief visit to Ferrara where he was awarded a doctorate in canon law. Evidently, Ferrara undercut other universities' examination fees and would examine students who had taken courses elsewhere as an easy way to make money. Copernicus returned home therefore officially qualified in canon law and partially trained as a physician. He was also, by common consent of those who knew him as one of the most cultivated minds in Europe, well read in Ancient Greek philosophy and already respected as a formidable mathematician and expert astronomer.

From his return, Copernicus was kept busy as secretary, translator and personal physician to his uncle, the Bishop. When his uncle died in 1512, Copernicus moved to Frombork and became engaged in the demanding administrative duties of a cathedral canon. Among other things, he was involved in drawing up legal agreements with tenants, settling disputes, allocating rents, assessing and awarding poor-relief and pensions, surveying and valuation, managing public health, and even dealing with a currency crisis. Even so, he also managed to write the first statement of his new astronomical scheme, a 20-page outline known as the *Commentariolus* (*Little Commentary*). Written in about 1512, it was circulated in manuscript form among astronomical colleagues and a few other learned men.

The following year, Copernicus was invited to participate in a collaborative effort to reform the calendar that had been proposed by a session of the Fifth Lateran Council, one of an important series of major Church councils. Unfortunately, Copernicus's reply has never been found, but he does not seem to have taken part in these efforts (which did not bear fruit anyway; the reformed Gregorian calendar, under Pope Gregory XIII, was not introduced until 1582). It is possible, in view of Copernicus's emphasis on cosmology rather than the merely practical art of astronomy, that his reply urged the need to reform astronomy as a whole, as opposed to simply trying to determine the precise length of the year and the lunar month. It's usually assumed, anyway, that he began work on his *On the Revolutions of the Heavenly Spheres* shortly after this invitation.

Copernicus struggled to write his great book amid all the ongoing distractions of his life as a canon. Sometimes he had to lay the book aside for two years or more. There was a pressing distraction in 1520, for example, when Copernicus had to take charge of preparing the town of Olsztyn for siege. Throughout this period, he spent all his time gathering every arquebus (a type of portable gun) in the district and trying to buy another 50 more.

Although progress on his big book was necessarily slow, thanks to the *Little Commentary* Copernicus's astronomical ideas were gradually spreading abroad. In 1533, Pope Clement VII was so impressed with a brief account of Copernicus's ideas that he rewarded the messenger, the Austrian chancellor Johann Albrecht Widmanstadt, with a valuable Greek manuscript. When Cardinal Nicolaus Schoenberg came to hear of Copernicus's ideas in 1536, he urged him to publish as soon as possible. The book was all but complete in 1539 when Georg Rheticus, a professor of mathematics at the University of Wittenberg who had heard of Copernicus's theories, came to visit.

Rheticus immediately became a devoted disciple of the now-elderly astronomer, and stayed with him for three years. The younger man's enthusiasm no doubt helped Copernicus to overcome the final hurdles that prevented him from completing his work. It used to be believed that Copernicus completed his book in about 1531 but decided not to publish it until the end of his life because he was afraid of the reaction it might cause. More recently, however, historians with a full awareness of the administrative burdens imposed by his Frombork canonry, and with an expertise in technical astronomy allowing them to form a realistic idea of what Copernicus was up against, have concluded that he could hardly have finished the book much earlier than 1543. It would seem that Copernicus's education and formidable intellect meant that his talents were continually being drawn on by his fellow creatures who needed his financial, administrative, political, legal and medical expertise far more than they needed his astronomical skills.

If Copernicus was anxious about the possible reaction that his crazy idea might cause, Rheticus helped to put these ideas to rest. He asked Copernicus if he could publish a brief announcement and description of Copernicus's system, and the old man agreed. This *First Account*, in which Rheticus declared Copernicus to be the equal of Ptolemy, appeared in 1540. No great outrage ensued. So Copernicus completed the final numerical revisions his calculations required and prepared his manuscript for the printing press.

On the Revolutions of the Heavenly Spheres was to be printed by Johann Petreius, a distinguished printer in Nuremberg with valuable experience in printing technical astronomical works. Rheticus took the manuscript to Petreius in May 1542 and remained in Nuremberg to check the proofs as they came off the presses. In those days, of course, any couriers who might have been used to carry the proofs back and forth from Nuremberg to Frombork were much slower than couriers today, and even less reliable (if that's possible). So, the manuscript was now completely out of Copernicus's hands and he had to rely on Rheticus to make sure everything was all right. Unfortunately, Rheticus had to leave Nuremberg in October to take up a new job as a professor at Leipzig. Supervision of the printing was now taken over by a friend of Rheticus's and a correspondent of Copernicus's, a man called Andreas Osiander (1498–1552).

Hereby hangs a tale. If Copernicus was to be the saviour of astronomy and cosmology, Andreas Osiander was to be his Judas Iscariot. Osiander added a preface at the beginning of the book, in front of Copernicus's own preface. When they saw this unauthorised preface Copernicus's two closest friends, Rheticus and Tiedeman Giese, Bishop of Kulm, were outraged at what they saw as Osiander's betrayal of their distinguished old friend, and tried to make Petreius produce a corrected edition. This lawsuit failed, however, and the book was distributed with the false preface. Osiander had left his preface unsigned and so throughout Europe it was generally assumed that it had been written by Copernicus himself. What did this preface say? In a nutshell, it said that the ideas in the book weren't true. Don't take these ideas literally, it suggested – of course the Earth isn't moving (the very idea!); just think of what is presented here as nothing more than a revised and handy way to calculate heavenly motions.

In December 1542, Copernicus suffered a massive stroke and was left paralysed. The book was finally ready for distribution to buyers and booksellers in March 1543, but it is said (by Tiedeman Giese, formerly a fellow canon at Frombork) that Copernicus himself did not receive a copy until May 24, the very day he died. This sounds like a finely tragic story dreamed up at a dinner party, but it might just be true. If it is true, it is all too easy to slip into romantic mode and suppose that Nicolaus Copernicus, proud restorer of the true cosmology, upon seeing the preface that brushed his cosmology aside and turned him into just another practitioner of the practical art of astronomy, had a second stroke, or, to remain in romantic mode, died of a broken heart.

Renaissance man

This brief account of Copernicus's life provides few clues as to why he did what he did and why he thought what he thought. Most of his life as a canon seems like a distraction rather than a stimulus to him. Besides, we know from the outline of his new system of cosmology in the *Little Commentary* that he had already developed this idea by 1512, only nine years after taking up his duties as a canon. The most likely source for inspiration seems to come from his student days when we know he first developed an expertise in astronomy, even though he was supposedly studying law. But expertise in astronomy was not sufficient to account for his great achievement. What made Copernicus stand out from the (admittedly small) crowd of other astronomers?

The answer must surely lie in his excitement about the rediscovery of Ancient wisdom that marked the Renaissance. We've already seen that one very important consequence of the Renaissance, at least among the liveliest intellects, was a new confidence in seeking to understand things for oneself. Adherence to the authority of Aristotle had never been entirely abject but it had tended to strongly restrict the terms of discussion. Disagreement from Aristotle amounted to little more than showing that his arguments weren't as conclusive as he implied; they very rarely led to a new way of thinking (remember Nicole Oresme's suggestion that the Earth could be moving).

Renaissance discoveries of philosophical works by other Ancient writers, however, showed that it was possible to offer radically different explanations of the same phenomena. In some cases such alternative philosophies won new followers among Renaissance scholars; there were revivals of Ancient atomism and of Stoicism, and above all of Platonism. But there was also a revival of Ancient scepticism, with its insistence that no philosophical argument could be trusted as a reliable guide to truth. Increasingly, therefore, philosophers were breaking away from the old doctrines supported by the authority of Aristotle and seeking to establish the truth about things by the direct study of nature.

This new Renaissance way of thinking was to have major repercussions elsewhere. There were breathtaking changes in the visual arts. The great Renaissance painters Leonardo (1452–1519), Michelangelo (1475–1564), Raphael (1483–1520) and others were able to represent things with improved realism – bodies with convincingly real anatomy, buildings that looked three-dimensional, landscapes where distant objects looked hazier than things close by. Such realistic improvements over the merely stylised and symbolic images of medieval paintings were achieved by artists paying greater attention to the close observation of nature.

Even the dramatic changes in religion that followed on from the formation of the Protestant churches after Martin Luther (1483–1546) denounced the corruption of the Roman Catholic Church and the Papacy, can be seen to stem from a rejection of authority. Certainly, Luther urged the faithful to follow the Scriptures, but this was the religious equivalent of urging them to study nature for themselves. Catholics were discouraged from reading the Bible, being told that to understand it required the authoritative guidance of a priest. Luther, rejecting many of the interpretations that priests were instructed to provide, insisted that everyone should study the Bible for themselves. Protestantism was to be a priesthood of all believers in which everyone was their own priest.

So the Renaissance provided the first opportunities for Western European intellectuals to break away from the traditional authorities. There was nobody earlier who came close to rejecting the dominant authority of Aristotle the way Renaissance thinkers did. The time was ripe, therefore, for a brilliant mathematician like Copernicus to exploit his skill in a way that no mathematician could have done before. But great innovations are not made by rejecting all that has gone before and attempting to start completely from scratch. This would be like trying to lift yourself up by your own shoelaces. The great Renaissance painters rejected the emblematic approach of medieval painting, but they used the natural world to provide their models. Protestants rejected the authority of the Pope, but they built anew on the foundation of the Scriptures. Similarly, we can discern alternatives to Aristotle's authority, which seem to have played a role in stimulating Copernicus's new approach.

Foremost among such stimuli must have been the newly discovered works of Plato. Throughout the Middle Ages, only a couple of fragments from Plato's writings were available. It was known, however, that one of the greatest of the Early Fathers of the Church, St. Augustine (AD 354–430), had been a great admirer of Plato before his conversion to Christianity, and had admitted in his autobiographical *Confessions* that if he'd come across Plato's writings after his conversion they might have lured him away from his faith.

There was considerable excitement, therefore, when it was discovered that all of Plato's works were known to scholars in the Greek-speaking territories of the Eastern Orthodox Church (which had split away from Roman Catholicism over the period from about 867 to 1204). What seemed so remarkable about Plato to Renaissance thinkers was that his philosophy seemed to prefigure Christian doctrines in so many respects. His ideas on the afterlife and the immortality of the soul seemed particularly close to Christian ideas, and some of his followers even developed ideas that seemed to provide pre-Christian confirmation of the truth of the doctrine of the Holy Trinity.

Suggestions of Christian doctrines among pagans who lived centuries before Christ seemed to confirm the traditional belief in the all-knowing wisdom of Adam before the Fall. Adam must have known about the Holy Trinity, the immortality of the soul and related ideas, and tried to pass them on. For the most part, fallen humankind was too blind for this knowledge to be preserved, it was assumed, but a good and wise man like Plato was still able to recognise such truths for what they were.

This view of the importance of Plato's philosophy was strongly reinforced by the discovery of another group of Ancient writings. This time they were the writings of Hermes, known as thrice-great, or Hermes Trismegistus. Hermes is the Greek god who is always depicted with winged sandals and a winged helmet, and holding a caduceus – a short staff (also winged, and entwined with two snakes). He is the only Greek god who was ever reputed to have written anything down. The disappointing truth, of course, is that these works were not written by Hermes Trismegistus. Whoever *did* write them signed Hermes' name on them, and the Renaissance scholars who discovered them believed they *were* written by him.

Hermes – whom Christian scholars did not believe to be a god but a supremely wise man

– was thought to have lived even earlier than Plato. Remarkably, his ideas were very like Plato's although they showed even more similarities to Christian doctrines. He, too, seemed to believe in the Holy Trinity, the immortality of the soul, punishment for sins in the afterlife and so on. Clearly, Hermes must have derived these ideas from one of Adam's close descendants. Renaissance scholars soon came to believe that alongside the Judaeo-Christian tradition, stemming from Moses, there was a pagan tradition stemming from Hermes Trismegistus.

Now, let's get things straight. Looking back, we can see why it is completely unsurprising that Plato's philosophy seems to prefigure Christian doctrines. Augustine was not the only Early Christian admirer of Plato. Virtually all the Early Fathers of the Church (the leading intellects of their day) were admirers of Platonism, which was undoubtedly the most cerebral, intellectually satisfying and also the most spiritual of Ancient systems of philosophy. Whether consciously or unconsciously, the Early Fathers incorporated many features of Platonism into their Christian theology, including the notions of the Trinity and the immortality of the soul. Small wonder that Renaissance writers discovering Plato, after being raised on a Platonised Christian theology, should notice similarities between his thought and their Christian beliefs. As for the writings attributed to Hermes Trismegistus, they were actually written by Platonists living in the early Christian era, in the 2nd or 3rd centuries AD, who incorporated Christian ideas with their Platonism! Again, it's no wonder that a Renaissance scholar should find the writings of Hermes even closer to Christianity than those of Plato.

Anyway, the discovery of what the leading scholars of Europe took to be an Ancient tradition of pagan thought from Hermes to Plato, which because of its similarities to Christianity must have originated with Adam, proved to be immensely exciting to Renaissance intellectuals. It meant that all aspects of Plato's philosophy were considered with the utmost seriousness.

We know that Copernicus learned Greek in order to be able to read Plato's writings in their original language. This same tradition, which saw Plato as a Pythagorean, would also have stimulated Copernicus's interest in Ancient Pythagoreanism. Copernicus could hardly have failed, therefore, to learn about the Ancient emphasis on cosmology.

There was another aspect of this Renaissance admiration for the Hermetic and Platonic tradition that Copernicus was able to exploit in his work. The first translator of the writings of Hermes and Plato into Latin, thus making them available to all scholars (all educated men could read Latin, but few could read Greek), was an Italian philosopher called Marsilio Ficino (1433–99). The Hermetic writings, although based on Plato's philosophy, were essentially mystical and magical in their focus, and Ficino tended to read Plato's original philosophy in a magical way too (remember, Ficino thought the Hermetic writings came first, and Plato followed). Ficino developed his own magical philosophy out of this mixture, which, among other things, represented the Sun as the visible analogue of God in the universe. God, after all, began his creation by letting there be light, and the Sun is the chief source of light in the universe and the ultimate source of all life on Earth. These ideas proved to be very influential and Copernicus didn't hesitate to draw on them. By way of circumstantial support for his theory, Copernicus was able to write in Book I, Chapter 10 of *On the Revolutions of the Heavenly Spheres*:

At rest in the middle of everything is the Sun. For, in this most beautiful temple, who would place this lamp in another or better position than that from which it can illuminate the whole thing at once? For, the Sun is not inappropriately called by some the lantern of the universe, by others its mind, and by yet others its ruler. Hermes Trismegistus designates it a visible god, and Sophocles' Electra, the all-seeing. So indeed, as though seated on a royal throne, the Sun governs the family of planets revolving around it.

Historians have argued about the precise importance of Copernicus's knowledge of Pythagorean theories in which the Earth was considered to be moving. Although Copernicus himself suggests in his preface to *On the Revolutions* that he got the idea as a result of 'rereading the works of all the philosophers which I could obtain to learn whether anyone had ever proposed other motions of the universe's spheres', most historians take this with a pinch of salt. As we saw earlier, if Copernicus was to have a hope of persuading his contemporaries of the truth of his new system, he had to be able to show that it was in fact old (preferably as old as Adam).

It seems quite likely, however, that he actually stumbled across the idea while trying to work out a way of accounting for planetary movements without needing to use the Ptolemaic equant point. It was at this point that he was in a position to realise that by putting the Earth in motion around the Sun he had found a way to unite astronomy with cosmology once more. It was possible in the Ptolemaic system for the planets to be in any order. Ptolemy in fact had opted to put Mercury just beyond the Moon, Venus outside Mercury and the Sun next. So Venus was closer to the Sun than Mercury (which is wrong). He put the outer planets in the right order simply by assuming that the slower its motion around the sky the further away it was, but in Ptolemy's system there was no geometrical reason why they had to be in that order.

Compare this with the heavenly situation as Copernicus now saw it. The geometry of Copernicus's system allowed him to calculate the relative distances of the planets from the Sun. Only one order was geometrically possible. Simultaneously, Copernicus could establish the periods of the planet's revolutions around the Sun. For the outer planets this was the same as Ptolemy had said, but Ptolemy had had to suppose that Mercury and Venus, which were never very far from the Sun, shared the Sun's annual period around the Earth. Copernicus was able to establish that Mercury circled the Sun in about 88 days, Venus in about nine months. The order, Mercury, Venus, Earth, Mars, Jupiter, Saturn, determined by period of revolution around the Sun, tallied perfectly with the order established by geometry. Copernicus's astronomy established its own unique cosmology.

This must have immediately struck Copernicus as what his beloved Plato had in mind when he challenged the astronomers of his day to 'save the phenomena', by showing how the seeming complexities of heavenly movements are compatible with uniform circular motions and with a plausible physical reality. This is not to say that Copernicus's scheme is Platonist because it puts the Earth in motion about a central Sun; Plato, just as strongly as Aristotle, assumed the Earth was at the centre of the system. What *is* Platonic about it is its emphasis on an astronomy that entails and confirms a unique companion cosmology. This is what Eudoxus had tried to do and, no doubt, what Aristotle thought Eudoxus had done when he adapted the Eudoxan system in his own philosophy. However, as we have seen, by Ptolemy's time technical astronomy and the received cosmology had drifted apart, neither being quite compatible with the other.

Not every astronomer was in a position to see this advantage in the Copernican system, nor to set such great store by it. A second Copernicus would no doubt have had the mathematical ability to see how a moving Earth provided a fixed order of the planets, but he might well have continued to think that such a scheme must be false because it is incompatible with Aristotelian physics and common sense, to say nothing of religious traditions. Copernicus's discovery could only have been made and recognised for what it was by someone who was thoroughly steeped in the Platonic emphasis on the importance of having an astronomy that served cosmology.

In what was probably his most influential book, *Platonic Theology*, Marsilio Ficino had suggested that only someone who possessed 'the same genius as the Author of the heavens' (by which he meant the same genius as God) could see 'the order in the heavens, its progressions and proportions'. Copernicus himself, right at the beginning of his book, reminded his readers that Plato had said that only someone with correct knowledge of the Sun, Moon and other heavenly bodies 'can become, and be called, godlike'. Copernicus must have felt himself godlike when he realised that his astronomy provided a coherent order for the heavens in a way that no previous astronomy had. As he wrote in Book I, Chapter 10 of *On the Revolutions of the Heavenly Spheres* a little later on:

In this arrangement, therefore, we discover a marvellous symmetry of the universe, and an established harmonious linkage between the motion of the spheres and their size, such as can be found in no other way.

Renaissance mathematician

There was another important aspect of the Renaissance that helps us to understand how Copernicus could believe the Earth was in motion simply because the geometry of his system seemed to suggest it was. It was just at this time that the status of mathematics and mathematicians was changing. As we've seen, mathematics had been relegated to an inferior position during the Middle

Ages, thanks to Aristotle's insistence that it couldn't really *explain* anything (which could only be done by talking about reasons and causes). This was beginning to change in the Renaissance, and mathematics was being valued more and more.

Nothing is ever simple and the story of the rise in importance of mathematics is made up of many interwoven threads, but one aspect of it is bound up with an increasing concern with practical utility. The Renaissance was not just a period when scholarship and the fine arts began to flourish as never before. It also saw the beginning of the end of the old feudal system, the rise of mercantilism and the invention of banking. This was the period of European exploration and the first attempts at colonisation. Three great inventions changed the nature of the world – the magnetic compass made it possible for ships to strike out across the open oceans; gunpowder was to change the face of battle; and the printing press was to change the nature of information exchange.

Each of these factors interacted with and was affected by the others, of course, as well as with the other characteristics of the Renaissance we've already mentioned – the rejection of authority and the rise of independent discovery, the rejection of Papal authority and the establishment of various Protestant Churches, and so on. One other important element entwined with all these things was an increased perception that knowledge should be put to some use for the benefit of mankind, or (alas, more usually) for some privileged part of it.

Some intellectual pursuits had always been intended for practical benefits, of course. Alchemy, astrology and other magical arts were the most obvious ones, but the mathematical arts or sciences came close behind (indeed, throughout the Middle Ages the mathematical subjects were often included with magic). By contrast, the study of Aristotelian natural philosophy, as it was pursued at the universities, was never justified in terms of its practical usefulness. The purpose of such élite philosophy was to promote understanding of the world and how it worked. If it was acknowledged to have any practical benefits, those benefits concerned its usefulness in bolstering religion. A greater understanding of the world helps us to acknowledge the omnipotence and benevolence of the supreme Creator, or so it was said. But increasing dissatisfaction with Aristotelian natural philosophy, together with the rise in concern for knowledge with real practical benefits, led to a reassessment of the role of mathematical knowledge.

First manifesting itself as an increasing focus on basic mathematical techniques in elementary schools, useful techniques were frequently taken up by craftsmen. One of the most spectacular of these developments was the use of geometrical perspective to give pictures the illusion of three-dimensionality. Those painters who could learn and exploit this technique soon became much fêted and, as is well known, the best of them rose dramatically in social prestige. It was in the Renaissance that the great painter began to be seen for the first time as a 'genius', worthy of admiration by even the highest ranks in society. The development of algebra allowed the solution of previously impossible problems, and mathematicians were quick to show how such a powerful problem-solving tool could be used in everyday life. Meanwhile, rapid changes in everyday life itself, brought about by the wider changes of the Renaissance, led to increased demands for surveyors, military engineers, navigators, cartographers and the like. The result was that mathematicians came to be increasingly admired, and to hold themselves in higher intellectual esteem.

This, in turn, led to a reassessment of the intellectual status of mathematics among intellectuals. There was a radically new tendency towards allowing mathematical analyses of a situation to count as equally enlightening as philosophical explanations. Where previously a mathematical account was considered to be inadequate because it did not show how the events in question were physically caused (whether the behaviour of light rays, the motion of stars or the increase in a capital sum due to interest), it could now be supposed that the mathematics itself suggested its own kind of cause. A geometrical proof of Pythagoras's (6th century BC) theorem, for example, certainly explained why the square on the hypotenuse of a triangle equalled the sum of the squares on the other two sides, so why couldn't a geometrical account of the behaviour of light rays, say, be said to explain why a magnifying glass can be made to set fires? Recognising the increased power of mathematics, more and more philosophers were willing to extend the relevance of purely geometrical 'explanations' to applied mathematical subjects like optics and astronomy.

Copernicus was to take this to its logical conclusion. None of the Renaissance philosophers who discussed this issue before Copernicus's ideas became known could have guessed where these ideas would lead. They no doubt had in mind such puzzles as why a lens made things look as if they were upside down, why a stick looks bent in water or why Mars looks larger and brighter when its motion across the sky is retrograde. Copernicus extended the same kind of answer to the inevitable question raised by his new system of astronomy: why does the Earth move? The answer is in the geometry of the system. But this was a huge leap.

To a large extent, as we'll see, this kind of answer was still considered to be completely unacceptable by almost everyone. To say that the Earth moves because the geometry of the heavens suggests it must was highly controversial. The point is, though, that not so very long before Copernicus, this kind of claim would have been completely unthinkable. Now, thanks to Renaissance reassessments of the nature and status of mathematical knowledge, it had at least begun to become thinkable. And Copernicus thought it.

• Chapter 4 •

What Was the Reaction?

Small beginnings

The publication of Copernicus's *On the Revolutions of the Heavenly Spheres* did not make much of a splash, although it did send out ripples that were eventually to have a major effect, directly or indirectly, on the whole of Western culture.

There were two main reasons for the book's initial lack of impact. The first was Osiander's preface, which told readers not to take seriously the suggestion that the Earth is moving. The second was the pretty obvious fact that it was a book that very few people could read. The first ten chapters are fairly easy going, provided you can believe what you are reading (which, of course, would have been difficult for people at the time). It's here that Copernicus provides an outline description of his system – stationary Sun at the centre, moving Earth, the works. After that, you need to be pretty hot on Ptolemaic astronomy and the sort of person who is undaunted by a bit – no, a lot – of advanced spherical geometry. In short, you need to be an astronomer.

Copernicus and the astronomers

Few astronomers were immediate converts to Copernicus's system. Historians have tried to find card-carrying Copernicans in the years after the publication of his great book. So far – and it looks as though this Europe-wide survey is now complete – they've managed to find about a dozen in the years from 1543 to 1600. But this doesn't mean that Copernicus was ignored by all the other astronomers. Far from it. They all knew a good thing when they saw it and adopted many of his ways of calculating heavenly movements.

When the astronomer Erasmus Reinhold (1511–53) employed Copernican methods to calculate a new set of astronomical tables, all astronomers used them gratefully. These tables, known as *Ephemerides*, tabulated the daily positions of the Sun, Moon and planets for a year as seen from a given location. These could then be used as the basis for calendrical, astrological or navigational calculations without individuals having to make their own observations and do their own calculations from scratch. Astronomers were also quick to recognise the superiority of Copernicus's model for calculating the movements of the Moon and his account of the precession of the equinoxes. But all of these things could be used, in an entirely pragmatic way, without having to believe that the Earth was in motion.

This response marks the success of the preface that Andreas Osiander inserted right at the beginning of *On the Revolutions of the Heavenly Spheres* in what Copernicus's close friends Rheticus and Tiedeman Giese saw as an act of betrayal against him. Although it is obvious from Copernicus's own preface, and the opening chapters of the book, that he really *did* believe in the motion of the Earth, Osiander's preface made it possible to dismiss this idea and pretend it wasn't significant. 'The author of this work has done nothing blameworthy', this unsigned preface read,

> *… for it is the duty of an astronomer to compose the history of the celestial motions through careful and expert study. Then he must conceive and devise the causes of these motions or hypotheses about them. Since he cannot in any way attain to the true causes, he will adopt whatever suppositions enable the motions to be computed correctly from the principles of geometry for the future as well as for the past. The present author has performed both these duties excellently. For these hypotheses need not be true nor even probable.*

We saw earlier that natural philosophy was held by Aristotle and his followers as the only true science because it provided explanations based on causes, while astronomy provided only geometrical descriptions. Osiander emphasised these conservative assumptions: 'For this art,' he wrote in his preface (and notice he called it an 'art'),

> ... *is completely and absolutely ignorant of the causes of the apparent non-uniform motions. And if any causes are devised by the imagination, as indeed very many are, they are not put forward to convince anyone that they are true, but merely to provide a reliable basis for computation.*

He finished his unauthorised preface with a flourish:

> *Let nobody expect anything certain from astronomy, which cannot furnish it, lest he accept as the truth ideas conceived for another purpose, and depart from this study a greater fool than when he entered it.*

Here, then, was Copernicus's life's work dismissed as merely a handy way of calculating heavenly motions for practical purposes, but not intended to be a true account of the universe. What Osiander had done was to emphasise in extreme terms the nature of the medieval stand-off between Aristotelian cosmology and Ptolemaic astronomy. Although earlier astronomers had never seen their role as so far removed from establishing the truth as Osiander presented it, what he wrote still seemed to make sense; it could be recognised by contemporary readers as a starkly exaggerated summation of the accepted difference between cosmology (which was concerned with how things really were) and astronomy (which was concerned only with being able to calculate heavenly motions). Thanks to Osiander, therefore, it was possible to use Copernicus's astronomical scheme, in whole or in part, without being a Copernican.

Besides, there were two other problems with the Copernican theory as far as astronomers were concerned. For one thing, it wasn't really all that much more successful than the Ptolemaic system in calculating celestial motions. Many of the details of the models for individual spheres, especially the Moon and Mercury, were so complicated that the system was just as difficult to use as the Ptolemaic system. What we can see, with hindsight, is that Copernicus's system was bound to be seriously flawed because he was still assuming, as the Greeks did, that the heavens must be composed of perfect spheres, rotating uniformly about their centres. Accordingly, he still had to use epicycles and eccentrics, and because he refused to use the equant he had to make his eccentrics and epicycles do extra work.

The fixed stars presented another major astronomical problem for Copernicanism. If he was right, the Earth's motion around the Sun meant that over any period of six months its position varied by 2,284 times the radius of the Earth. Since the radius of the Earth was known to be about 25,000 miles, this was a huge distance, and this ought to mean that stellar parallax should show up. 'Parallax' is the name given to the fairly familiar phenomenon of things seeming to change position when viewed from different places. As you walk along a straight road on a moonlit night you'll notice the lamp-posts, trees and houses dropping back as you move past them. You soon find yourself having to look sideways to see a lamp-post that was once straight ahead of you, and shortly after you have to look back over your shoulder to see it.

But what about the Moon? If you start your walk with the Moon on your right, throughout your walk it will always appear to be in the same position relative to you, to your right. It will seem to be following you down the road, while lamp-posts and houses move backwards away from you. A fancy way of describing this is to say that the Moon has no parallax, but the lamp-posts do. The difference depends upon the distances involved. Objects further away show less parallax. The Moon is so far away that you can't detect parallax as you stroll down a street (although astronomers can detect the Moon's parallax).

Let's switch to a bigger picture. The Earth takes a

stroll around the Sun of what Copernicus thought was about 57 million miles (but which is closer to 200 million), and yet we always have to look in the same direction to see a particular star. You'd imagine that if you had to look east to see a particular star in winter, then in summer, when the Earth is on the opposite side of the Sun, you might have to look west to see the same star, but you don't. The star is still over on the east (see Figure 7).

This can only mean two things. Either the stars are an unimaginably long way off – much, much further away than estimates of the size of the Aristotelian universe suggested – or the Earth never does take strolls around the Sun. You can guess which was the most favoured option. But the small handful of card-carrying Copernicans had to commit themselves to a vastly expanded sphere of the fixed stars.

This highly technical detail was to have the most far-reaching consequences. You might even say infinitely far-reaching. The earliest Copernicans began to develop the idea that the universe is infinite in size and that the stars, instead of being luminous points on a celestial sphere surrounding the Earth, are scattered throughout an infinite space. Mighty oaks from little acorns grow, and our concept of an infinite universe grew from Copernicus's picture of the solar system.

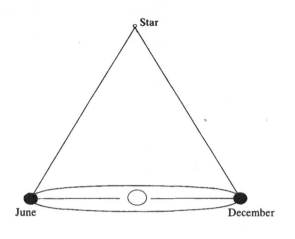

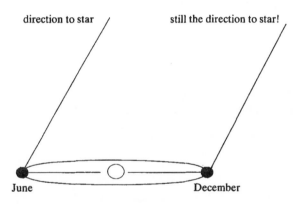

Figure 7. The top picture shows how you might expect to look at the same star after an interval of six months, if Copernicus is right, and the Earth moves around the Sun. The bottom picture shows what actually happens. You still have to look in the same direction (the star shows no parallax). This means that either the stars are unbelievably far removed from the Earth, or Copernicus is wrong, and the Earth remains in the same place.

The concept of an infinite universe was to have implications way beyond astronomy. The Aristotelian world picture had been combined with the Christian world picture in the 13th century, and since then it had always been assumed that God and all the heavenly host lived above the sphere of the fixed stars (see Figure 8). Thomas Digges (*c.* 1546–95), an English astrologer who was one of the first to take the Copernican theory beyond the vault of heaven, simply assumed that God, the angels and the grateful dead lived up there *among* the stars, instead of *beyond* them.

It would require a book in its own right to trace the development of this idea, but by the end of the 17th century the idea that the solar system was surrounded by infinite empty space was generally accepted. The location of the theological Heaven now became a topic of dispute among theologians (as did the site of Hell, which some proposed was in the Sun, its fire continually being stoked by the souls of the damned), as growing ranks of sceptics and atheists used the empty universe of physics to deny the validity of religion.

It's ironic that most of the great natural philosophers or scientists of these times were devout religious believers who saw their studies of the physical world as a way of honouring God and his Creation, and yet their science was repeatedly and enthusiastically taken up by atheists. The theory

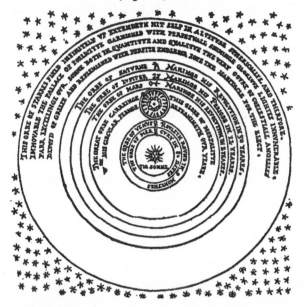

Figure 8. Illustration of the Copernican Universe from a book on astrology by Thomas Digges published in 1576, one of the earliest Copernican publications. Digges clearly shows the stars scattered throughout space, rather than on a sphere at a fixed distance from Earth. 'This orb of stars fixed infinitely up', he writes, 'extends itself in altitude spherically'. Even so, he goes on to say this is still the dwelling place of God, the angels and the elect.

of the infinite universe formed the background to the famous comment by the extremely devout mathematician Blaise Pascal (1623–62): 'The eternal silence of these infinite empty spaces frightens me.'

There can be no doubt that the infinity of space, hinted at by Copernicanism, was to play a major part in the decline of religion and the secularisation of the world picture. It would seem that Pascal, who later in life abandoned science and devoted himself to religion, was right to warn his Church to be afraid.

Copernicus and the Churches

To begin with, the Roman Catholic Church was not in the least bit afraid. The only known formal objection by a Catholic theologian was discovered in 1975, where it had been sitting on the shelf of a monastery library in Florence ever since it had been written in 1544. It just so happened that the author, Giovanni Maria Tolosani (1470–1549), had astronomical as well as theological expertise – he attended the Church Council that had invited Copernicus to help in its plan to reform the calendar. His objections to Copernicus did not arise from his astronomy, however, but from his presumed deficiency in physics. Tolosani emphasised the Aristotelian arguments about the impossibility of a moving Earth (which we'll come to soon) and insisted that, because Copernicus tried to justify the motion of the Earth merely on mathematical grounds, he clearly didn't know 'how to distinguish between true and false' or what were the proper 'modes of argumentation', which were the preserves of philosophy, not mathematics.

Tolosani obviously had a bee in his bonnet about this, but there's no evidence that anybody else in the hierarchy of the Catholic Church was bothered in the least. It seems likely that Osiander's preface to Copernicus's great work did precisely the job Osiander wanted it to do; by suggesting the theory wasn't intended to be taken as true, but merely as a means of making astronomical calculations, most Churchmen decided it was of no concern to them.

It was only when the Copernican theory began to be vociferously proclaimed as physically true that the Catholic Church began, reluctantly, to pay attention. The first inkling Church leaders might have had of the potential danger appeared in the writings of a renegade Dominican monk, Giordano Bruno (1548–1600). Bruno believed that Christianity was corrupt and that the true religion was the religion of the Ancient Egyptians. Members of the Inquisition felt they had little choice but to burn him at the stake for heresy, which they did in 1600. Unfortunately, Bruno also used Copernican astronomy to advocate an infinite universe, which he saw as a more fitting creation of an infinite deity, and suggested there must be life on other worlds. Copernicus may have been tainted by association, but the Church made no formal ruling about Copernicanism at this time. Osiander's stratagem was still holding.

This all changed thanks to a tactless and pugnacious self-publicist by the name of Galileo Galilei who had a talent for making enemies. Galileo brought it to educated attention all over Europe that the Copernican theory was not just a way of working out planetary movements but a theory of cosmology that claimed the Earth really and truly is moving. Galileo produced some extremely exciting and highly suggestive evidence in favour of Copernicanism by turning the newly invented telescope to the heavens, and at first Church leaders were as delighted and impressed as everyone else. But, beleaguered by enemies, Galileo embarrassed the Church by taking the dispute into the realms of Biblical interpretation.

There are a number of places in the Bible where it is strongly implied that the Earth is stationary and the Sun is in motion. Galileo took the standard line that the Bible was written in such a way as to not baffle the uneducated, and thus spoke of heavenly movements in terms of the appearances. The Bible is meant to tell us how to go to heaven, not how the heavens go, he lightheartedly quipped. Unfortunately, he also insisted that if the Bible seems to contradict mathematical science, then the Bible must be reinterpreted to make it conform to mathematics.

Galileo was a mathematician – and one who, like Copernicus, regarded mathematics as superior by virtue of its certainty to philosophy. The

Church authorities, however, were still tied to the old Aristotelian ways of thinking in which mathematics was an inferior discipline that couldn't *explain* how things are. They did not see things Galileo's way, therefore.

The Catholic Church first ruled against the Copernican theory in 1616, but the main force of this ruling was effectively to confirm the attitude formulated by Osiander that this theory was merely a means of calculating heavenly movements and shouldn't be regarded as physical truth. Galileo's name was kept out of it at this time to protect his reputation. When he *was* arrested and condemned in 1633, after publishing a defence of the reality of the theory, it was the end result of a series of highly specific and personal circumstances. But this deserves a book of its own, and we can just say here that that's another story. The point is that, regardless of whatever modern atheists or positivistic scientists would like you to think, the Galileo case cannot be used to support a general claim that science and religion don't mix. True, they didn't mix in this case, but a close look at the history makes it clear why, and it was not because the Catholic Church, much less Christianity itself, was opposed in principle to scientific knowledge.

The Protestant Churches did not have a Galileo to deal with and so it is much easier to see how the alliance between science and Christianity usually worked. Protestantism began, after Martin Luther publicly rebelled against the corruption of the Catholic Church in 1517, in an atmosphere of anti-intellectualism, but this was soon countered by educational reforms introduced by Luther's right-hand man, Philip Melanchthon (1497–1560).

Melanchthon was one of those Renaissance thinkers, like Copernicus himself, who recognised the intellectual power of mathematics, particularly its certainty and its potential usefulness for everyday life, but also its value as a means of instilling discipline of thought and good habits of reasoning in students. Accordingly, the mathematical subjects played an enhanced role in Lutheran universities. Astronomy was regarded as especially useful because an understanding of 'the order of the heavenly motions' revealed the omnipotence and omniscience of God.

The University of Wittenberg, where Luther and Melanchthon were both professors, developed a strong tradition in astronomy, and it was from here, of course, that Georg Rheticus was to seek out Copernicus. Thanks to Rheticus's *First Account* of the Copernican theory, and his more personal advocacy of the new theory, the Wittenberg group quickly became familiar with the theory. Erasmus Reinhold, who made the first Copernican *Ephemerides*, or tables of astronomical positions, was also a professor there.

However, the Wittenberg group tended to be fairly cautious about Copernicus's ideas. All too aware of the Aristotelian objections to the physical possibility of a moving Earth and of its incompatibility with a few statements in the Bible (consider, for example, *Joshua* 10, verses 12–13 and *Ecclesiastes* 1, verses 4–5), the astronomers tended to take the line suggested by Osiander. They warmly embraced Copernicus's method of disposing with equants, his way of dealing with the complexities of the Moon's motion and other useful aspects of the theory, but remained absolutely non-committal on the underlying cosmology. The exception was Rheticus, of course, but he was regarded as a young hot-head and a bit over-enthusiastic.

Nevertheless, out of this Lutheran tradition of astronomical teaching were to appear, in the next generation, at least three other Copernicans. The greatest of these, Johannes Kepler (1571–1630), was certainly led to Copernicanism, at least in part, by his religious convictions. What's more, these same religious convictions were to lead him eventually beyond the Copernican system to discover the true structure of the solar system.

It was Kepler who first realised that the heavenly bodies do not move with uniform circular motions but, in fact, move on elliptical orbits with the Sun at one of the two geometrical foci of the ellipse. He also realised that the bodies do not move around these orbits at constant speed but that they speed up as they move closer to the Sun and slow down as they move towards the empty focus of the ellipse. Kepler even managed to establish the geometrical rule governing the speed of these motions (see Figure 9).

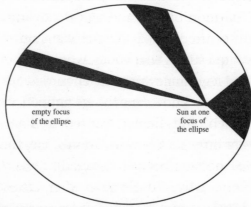

Figure 9. Kepler discovered that the planets actually move on ellipses, not circles. This is now called Kepler's First Law of Planetary Motion. He also discovered that the line joining the Sun to the planet sweeps out equal areas in equal times; this is now known as Kepler's Second Law. In the diagram the shaded areas are all equal, which means that the planet must move faster along its orbit when it is near the Sun than when it is further away. This ellipse is elongated for clarity. As the two foci of an ellipse are brought closer together, the ellipse approaches nearer to a circle (which can be regarded as an ellipse with both of its foci in the same place). The planetary orbits are closer to circles than this diagram makes it appear. It is also worth noting that Ptolemy's equant point is a close approximation to Kepler's Second Law. Seen from the empty focus, the planets would almost seem to be moving with constant speed.

But how could such precise astronomical discoveries, which are now recognised as scientific laws of planetary motion, have been inspired by Kepler's religious convictions? It's an interesting story.

Initially convinced, like Copernicus and Rheticus, by the harmonious cosmological advantages of the Earth-centred theory, Kepler went on to ask a couple of questions that today would not even be considered legitimate scientific enquiries. Scientists tend to want to know *how* things happen, not *why* they happen. But Kepler wanted to know why there were only six planets (Uranus, Neptune and Pluto had not yet been discovered). Since God is omnipotent, why did He stop after creating six planets? Presumably He could have created ten or a thousand, maybe even an infinite number.

Furthermore, now that the Copernican system enables us to establish exactly how far out from the Sun each of the planets is, the question arises: 'Why did God choose those particular distances?' Surely the obvious thing to do for a nice harmonious pattern would be to space them out equally, but God clearly didn't do that. The distribution seemed without rhyme or reason, but there had to be a pattern: God doesn't do things without a reason.

Kepler's first clue came one day while he was teaching his students about the conjunction of Jupiter and Saturn – that is to say, the moment when Jupiter and Saturn appear together at the same place. Typically for him, Kepler saw it as a revelation from God. He was drawing on the blackboard (some things never change) to show how these conjunctions move around the sky, forming a rotating equilateral triangle (see Figure 10). To his experienced eye, the circle formed at the centre by

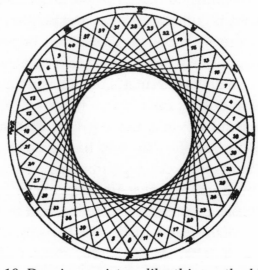

Figure 10. Drawing a picture like this on the blackboard for his students, showing how successive conjunctions of Jupiter and Saturn move around the sky (number 1 is at 3 o'clock, 2 at about 7 o'clock, 3 at about 10 o'clock, and 4 brings us back, nearly to our starting point, just above the number 1, and so on), Kepler noticed that the proportions between the outer circle and the inner circle were close to the proportions between the spheres of Saturn and Jupiter. This was to lead him to God's 'geometrical archetype'.

the rotating triangle and the outer circle seemed to be in the correct proportions to represent the outer sphere of Saturn and the sphere of Jupiter inside it. Maybe God had put the planets where He did by nesting geometrical shapes between the orbits.

Kepler began to test out this hunch by trying all sorts of different shapes between the different orbits. Eventually he cottoned on to the fact that he should be thinking in three dimensions, not two; the heavenly bodies were spheres not circles. And soon he had the answer. God had put the planets where he did by nesting them around the so-called 'Platonic solids'.

Start with the sphere of Mercury and put an octahedron around it so that the sphere touches the octahedron at the centre of each of its eight faces. The points of this octahedron now mark the place where the sphere of Venus should be. Now put an icosahedron around the sphere of Venus, touching it at its sides, and the vertices of that icosahedron mark the place where the sphere of the Earth goes. Then carry on, using the dodeca-hedron, the tetrahedron and the cube to mark out the spheres of Mars, Jupiter and Saturn. Then you have to stop (see Figure 11).

Here's the neat part. Even if you are God, you have to stop. The Platonic solids are those solids made up of exactly equal sides. The tetrahedron, octahedron and icosahedron of four, eight and 20

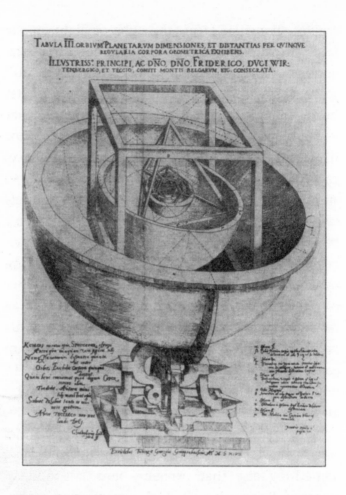

Figure 11. Frontispiece from Kepler's *Cosmographic Mystery*, published in 1596, in which he described what he believed to be God's geometrical archetype, or blueprint, for making the universe. The picture shows how the planetary spheres are nested alternately with the five Platonic solids, so providing the sizes of the spheres and limiting their number to six.

regular triangular faces, the cube of eight square faces and the dodecahedron of 12 regular pentagons are the only possible Platonic solids. You can't make a closed solid body by using only regular hexagons, or octagons or any other single, regular geometrical shape. The geometry just doesn't allow it. So, not even God, for all His omnipotence, can break the laws of geometry, and He has to stop at Saturn. God didn't create any more planets (remember Uranus, Neptune and Pluto hadn't been discovered, so they couldn't spoil the story), because he chose to space them out by nesting them with the Platonic solids. This is what Kepler called God's geometrical archetype. God chose to follow this blueprint, Kepler believed, to enable his human creatures to use the God-given gift of reason to discover the blueprint and so know, beyond all doubt, that God was out there. Kepler believed he was 'thinking God's thoughts after him'.

Now, you're probably thinking that Kepler must have been as nutty as a fruitcake. But he knew something else that probably you can scarcely imagine. Believe it or not, the geometrical archetype provided an astonishingly close fit. It wasn't perfect, but then neither were Copernicus's figures for the planetary distances. Kepler decided to take up a long-standing invitation he had received to work with a Danish astronomer called Tycho Brahe (1546–1601), who was engaged as the court astronomer to Rudolph II in Prague. At the time, Tycho was acknowledged to be the most accurate astronomical observer of all time. Kepler hoped that, with the benefit of Tycho's accurate astronomical data, he could establish beyond all doubt the truth of the geometrical archetype.

Tycho was not a Copernican. Indeed, he had undertaken the accurate observation of the heavens in the hope that a better reform of astronomy might be established by them. What he wanted to see was a system that succeeded in showing how astronomy and cosmology could be united, but without introducing an 'impossible absurdity' like the motion of the Earth. Tycho was a bit cagey with his observations, but he set Kepler to work on his data about Mars.

In 1609, nine years after moving to Prague, and by which time Tycho was dead, Kepler finally arrived at the elliptical orbit along which Mars moved with a regularly varying speed. Kepler said that he had worked so hard on his calculations during this period that 'he could have died ten times'. The important thing here is that a less religiously inspired astronomer than Kepler would never have persevered so long and so hard with the calculations; he would almost certainly have rested content with approximations based on combinations of circles rather than move to the unheard of notion that planets might move in ellipses. Before arriving at the correct shape of the orbit – the ellipse – Kepler had a mathematical model that fitted the observations closer than astronomers were used to. Most astronomers would have said it was close enough and triumphantly published their theory about how Mars moves. But not Kepler.

Kepler believed that Tycho Brahe had been sent by God to provide the accurate observations needed to reform astronomy, and to ignore the tiny discrepancy between his geometrical model and Tycho's observations was to spurn God's gift. The discrepancy amounted to eight minutes of arc (a minute in this context is a 60th of a degree), at a time when astronomers were happy if their calculations brought them to within 20 minutes of a recorded observation. Kepler went back to his calculations. 'Since it is not possible to ignore them', he wrote, 'those eight minutes point the road to a complete reformation of astronomy.'

Kepler's discovery of elliptical orbits certainly amounted to that, as did his realisation that the planets moved around those ellipses with varying speeds. Essentially, after more than a millennium in which it was assumed that planets must be moving uniformly in perfect circles, he had arrived at the right answer.

However, Kepler now found he had another problem – not one that could be considered scientific, but one that was very important to a man whose motivation was derived entirely from his religion. He wanted to know why God had chosen variable motions on an ellipse, when the most harmonious scheme seemed pretty obviously to be one in which the heavenly bodies moved uniformly around perfect circles.

To solve this problem, Kepler turned to the Ancient Pythagorean belief in the harmony of the spheres. According to this influential tradition, the order and harmony of the cosmos was such that the heavenly spheres made a celestial music as they turned. Kepler was quick to realise that what seemed like the most harmonious scheme of the heavens – uniform motions in circles – could only generate a monotone for each of the spheres. A heavenly body moving with a variable speed, however, could generate a range of musical notes, which, in combination with the notes of the other bodies, would result in complex polyphonic harmonies. What seemed to be the most harmonious system was, in fact, monotonous in comparison with the new cosmology discovered by Kepler.

In order to prove the truth of this idea Kepler now embarked on a comparison of the speeds of the planets at various positions along their orbits. Using Tycho's observational data and yet more mind-numbing mathematics, Kepler was able to show that the speeds of Saturn at perihelion (that is, closest to the Sun and moving at its fastest) and

Jupiter at aphelion (that is, furthest from the Sun and moving at its slowest) were in the ratio 1:2, which is precisely the ratio between the two tones when an octave is sounded. Similarly, the ratio between the speeds of Mars at perihelion and Earth at aphelion was 2:3, precisely the same agreement as that between two tones when a perfect fifth is sounded.

Similarly, by comparing the speeds of each of the planets at their aphelion and perihelion, Kepler was able to show that the two tones produced by Saturn at its slowest and its fastest were in the ratio of the major third; for Jupiter it was a minor third, for Mars the fifth, and so on for all the planets. Once again the agreement between Kepler's hunch and the physical reality proved to be astonishingly close.

Kepler now believed, not without good reason, that he had discovered what he called God's musical archetype. Referring back to his original idea of the geometrical archetype, Kepler concluded in Book V, Chapter 3 of *Harmonice Mundi* (*The Harmony of the World*, 1619):

It appears that the exact proportions of the distances of the planets from the Sun were not taken from the five regular solid figures alone: for the Creator does not depart from his archetype, the Creator being the true source of Geometry and, as Plato wrote, always engaged in the practice of Geometry.

So, God must have used both the geometrical and musical archetypes while creating the universe.

One extremely important spin-off from Kepler's studies of the speeds of planets at different parts of their orbits was what is now known as Kepler's Third Law of Planetary Motion. This enables us to work out the average distance of a planet from the Sun, provided we know the time it takes for it to complete a circle of the sky (its sidereal period). These periods have been known with great accuracy since the Ancient Greeks, so it was easy to establish the average distances from the Sun (the average between aphelion and perihelion) for each of the planets.

Kepler was now in a position to go back to the geometrical archetype to see if Tycho's data did reveal a better fit. Unbelievably, it very definitely and undeniably did! It was so close that it was well within the limits demanded by scientists today for the acceptance of new theories. In view of this, it's perhaps a good job that Uranus, then Neptune and Pluto have since been discovered to spoil Kepler's beautiful scheme, otherwise modern Western culture might all have been based on a belief in Kepler's God. After all, the truth of Kepler's Laws of Planetary Motion was finally established by another highly devout but idiosyncratic religious believer, motivated by his belief in God, and also convinced of the truth of Ancient Pythagorean wisdom. His name was Isaac Newton.

Speculation about what might have been can never be proven one way or another, of course. But what should be clear is that it would be wrong to assume that the Copernican theory was opposed by the Churches and dismissed by religious believers. Although Kepler's works were too mathematical for the common reader, in the long run his astronomical work did as much to prove the truth of Copernicus's Sun-centred, and hence *solar*, system as Galileo's more populist arguments. And, as we've seen, Kepler's religious beliefs did not hinder his achievement, they made it possible.

Copernicus and the Aristotelians

According to the principles of Aristotelian natural philosophy, the Copernican theory had to be nonsense. It wasn't simply a matter of the supposed motion of the Earth, although that certainly seemed impossible. It was also inconceivable that the Earth should be in the heavens. The Earth and the heavens were such totally different things, they had to be separate and distinct. You only have to look at the heavens to realise that they're like nothing on Earth, and vice versa.

Part of the strength of Aristotelian natural philosophy lay in the very fact that it was based on common sense and tended to reinforce basic intuitions. According to Aristotle, for example, heavy bodies fall faster than light ones. Although we now know this is false, we also know that it is what children instinctively assume to be true. Additionally, Aristotle believed, again as children do, that if something is moving there must be something moving it; motion doesn't carry on of its own accord. Once more, this is wrong. We now know that once something is set in motion it will continue moving in a straight line unless something interferes with it. This is how we can send tiny probes across millions of miles to Mars or Jupiter and beyond, and providing we can shoot them out beyond the Earth's gravitational pull, they carry on forever without the need of fuel-hungry rocket power. If Aristotle had been right, a probe to Mars would need to be pushed by rockets every inch of the way and would therefore need massive amounts of fuel. But, then again, if Aristotle had been right we would never have tried to send probes to Mars.

It's hardly surprising, therefore, that Aristotle should take it for granted that the Earth was stationary. Being a philosopher, though, he had to come up with the reasons why. This was easy. All 'earthy bodies', which essentially meant all heavy bodies, fall towards the centre of the world, and as this motion is perfectly natural and unforced it must mean that these earthy bodies are falling to their natural place, their home, if you will. It stands to reason, then, that the natural place of the whole Earth, the ultimate earthy body, is also at the centre of the world.

Now, it's easy for us to see the glaring flaw in this argument. We're used to the idea that the Moon and Mars and any other planets have their own gravity. We've all seen the film footage of Neil Armstrong and Buzz Aldrin jumping in their bulky suits in the reduced gravitational field of the Moon, and we know that the Earth is no different from anywhere else in the universe in having its own gravity. But we arrived at this knowledge as a result of Copernicus's work. Prior to Copernicus it wouldn't have been so easy to notice that Aristotle's argument about earthy bodies didn't have to apply to the Earth itself.

Ptolemy had reinforced Aristotle's point by asking his readers to imagine the Earth falling down, the way an apple does when it falls off a tree. Because the Earth is so heavy it would fall faster than anything else (he was an Aristotelian), so all the separate bodies on the Earth would be left behind 'floating on the air', while the Earth would fall 'out of the universe itself'. Clearly, this is a crazy idea. So the Earth can't be falling in the same way as other earthy bodies do, and must be at rest around the point that all the other bodies try to get to – namely, the centre of the world.

Connected with this Aristotelian theory of fall is a more general theory of motion. According to

Aristotle, motion can be of two basic kinds – natural and unnatural. Unnatural, or forced, motions don't concern us here, but natural motions do. These are motions that bodies spontaneously perform, according to their natures. Earthy bodies are heavy and fall towards the centre of the Earth. Fiery bodies are light and the sparks fly upwards, away from the centre.

Because Aristotle believed anything moving was continually moved by something, he had to assume that such spontaneous movements took place because something in the nature of the body made it move the way it did. This led him to conclude, also, that there must be a 'natural place' for the body, to which its natural tendency took it – earth to earth, fire to a sphere of fire, which was assumed to surround the Earth just below the sphere of the Moon.

Aristotle believed all bodies were composed of combinations of the four elements – earth, water, air and fire. Accordingly, he assumed that predominantly watery bodies would move to the sphere of water, which surrounded the sphere of the earth (except where the rigidity of the earth made it poke up above the sphere of the water to form land-masses), and airy bodies would move to the sphere of the air, between the spheres of water and fire.

These natural motions always took place in straight lines, either towards the centre of the world or away from it. But there was another kind of natural motion. Some bodies spontaneously moved in circles. Which ones? Heavenly bodies. So, this is how we know that the heavens are not made of earth, water, air and fire. If they were, they would have a natural tendency to move up or down. But they don't. Their natural tendency is to move in circles around the Earth. They must be made of something other than the four elements, a fifth element that is never found on Earth. If this fifth element accounts for the natural tendency of the heavens to move in circles, it must also account for the unchanging nature of the heavens. The four 'sub-lunar' elements are corruptible and change-able, as we are all too aware, but the fifth element must be perfect and incorruptible. Everything from the Moon outwards is perfect and unchanging with its own kind of unearthly physics, where circular motions are natural and there are no forced or unnatural motions.

This is why it made no sense to say the Earth is like the other planets. For Aristotelians it was a category mistake – like saying an elephant is 3 o'clock in the afternoon. The Earth is the Earth, and the heavens are the heavens, and they are very different kinds of things. You can't just put the Earth among the planets without disrupting the whole of natural philosophy. The only way you could make sense of a claim like that would be to show that the heavens are not made of a quintessence but that they are made of the same kind of thing as the Earth. However, if you were to claim that the heavens were made of earthy things, you'd then have to explain why they didn't all come crashing down to Earth, like other earthy bodies. No, it's no good Copernicus, it just won't work at all.

Besides, circular motion for the Earth is entirely unnatural. Any fool can see that the only natural motions here are in straight lines, up or down. If the Earth was made to rotate about its own axis it would be a forced motion that would result in massive cataclysm. Things would be hurled from their places and even the Earth itself would break up under the strain. Copernicus tried to answer this one, but in so doing he only highlighted the fact that he was failing (or pretending not) to see the point of Aristotelian physics. If the strain of a rapid rotation would break up the Earth, he said, it is much more likely to break up the sphere of the fixed stars, which is supposed to be rotating around the Earth once every 24 hours. The rotation of such a vast sphere in a day would have to be much faster than the rotation of the comparably tiny Earth in the same time. The Aristotelian would no doubt shake his head in despair at this suggestion. The circular rotation of the heavens is perfectly natural and would not put any strain on the aetherial sphere of the stars. The rotation of the Earth, however, would be a completely different kettle of fish.

Speaking of fish, how could they swim in the sea as they please if the Earth was in motion? How could the birds fly in the air? A bird that left its tree to fly west would never be able to fly back to its nest as it sped off on the eastwardly rotating Earth. If the bird left the nest to fly east it would be a painful experience. How far would it get before the tree caught up with it and batted it off into the distance? Cannons would have a different range when fired east or west, and if you shot one straight up in the air the cannonball would fall back to earth way to the west instead of endangering the lives of the gunners. None of these things happen, so the Earth must be stationary.

The Aristotelian principle that something moving must be moved by something also presented a big problem for Copernicus. If the Earth is continually moving, what is it that keeps it moving? There must be something, otherwise it would come to a complete halt. According to the Aristotelians, the heavenly spheres were moved either by 'intelligences', or angels, or by some innate power bestowed on them by God, so that it was in their nature to rotate ceaselessly.

We know that it is not in the nature of earthly substances to rotate ceaselessly, and the idea that the Earth could be moved by an angel did not seem plausible. It's one thing to suppose that an angel could maintain the turning on its axis of an aetherial sphere that is weightless and frictionless, but quite another to suppose that it could move a massive physical lump like the Earth not just around its axis but bodily through vast distances of space around the Sun.

Most of us no longer believe in angels and might imagine that, for those who believed in them, angels could do anything they liked, including moving the Earth. But nobody thought that. Only God could do anything. Angels were created by God, just like human beings, and thus were natural creatures with limited capabilities. Philosophers and theologians endlessly debated the abilities of angels, including whether they ever grew fatigued from turning the spheres. Angels were never considered as a likely source for the Earth's motion. If Copernicanism was to take hold, some other cause of the Earth's motion would have to be agreed on.

The Aristotelian world picture was the major stumbling block to an acceptance of the Copernican theory. Copernicus's fellow astronomers refused to accept its truth because they accepted Aristotelian natural philosophy and deferred to it as the discipline that explained physical truths. The religious authorities rejected Copernicanism because they had adopted an Aristotelian cosmology that shared the same common-sense view of the Earth's stability and the Sun's motion that could be read in the Bible.

Furthermore, Aristotle's claims about the fundamental dichotomy between the Earth and the heavens, in which the substances they were composed of and the kinds of motion and physical laws that applied in each realm were completely different, meant that it was almost impossible to envisage what Copernicus had in mind. The heavens were so completely and categorically different from the Earth that it made little sense to say the Earth was a part of the heavens and that the heavenly bodies were like the Earth. If Copernicus's astronomy was correct, it meant that both the earthly and the celestial physics of Aristotle would have to be rejected and replaced with a completely new unified physics.

Copernicus made little attempt to do this, however. *On the Revolutions of the Heavenly Spheres* addressed one or two of these problems in a rather unconvincing way, and ignored the rest. At one point he simply said that he was leaving the question 'to be discussed by the natural philosophers'. He was content, as we've seen, to rest his convictions on his mathematics. The geometry demanded a moving Earth and so it must be moving. But for most of his contemporaries, mathematics was not the sort of thing that could tell you truths about the physical world. The Copernican astronomy needed the support of Copernican natural philosophers.

Chapter 5

What Difference Did it Make?

A world of difference

In spite of the obliterating absurdity and evident impossibility of the Copernican theory, it gradually won more adherents, including some of the greatest minds of the age, and slowly but surely came to be accepted as true. As this happened, all the necessary changes in the world picture were also made – so that our picture of the cosmos today, where there are no heavens but only outer space, and where our universal physics is routinely confirmed by mathematics, grew out of the early efforts to establish the truth of the Copernican system.

As these scientific developments took place something else had to happen. The new science was no longer merely a formalised account of common-sense beliefs, the way it had been when Aristotle was *the* philosopher. The new science began by explaining how the Earth moved, a totally counter-intuitive notion, and went on to explain more and more natural phenomena in terms that bore little relation to everyday perceptions and assumptions. It would be easy to see this as a recipe for disaster – the new breed of natural philosophers becoming more and more wacky until they lost all credibility with the public. But, as we know, that's not what happened – although for a while, in the 17th century, the new science was indeed a frequent target for ridicule on the stage and in literature.

It wasn't long, however, before the new science proved so successful that to ridicule it became ridiculous. In spite of its counter-intuitive nature, therefore, the new science became increasingly authoritative. This was also the historical moment, of course, when the old authority of religion began to lose much of its former force. But the weakening of religious authority as the power of science increased was by no means a coincidence.

The Aristotelian world picture had become so firmly adopted by the Christian Church throughout the Middle Ages that its demise was truly shattering. The traditional world picture had a place for Hell and for Heaven. In between there was a hierarchy of creatures on Earth, from the vilest worm to men at the top (men were always at the top in those days). Above that was a hierarchy of heavenly beings, angels, archangels, principalities and powers (who turned the heavenly spheres), and five higher orders all the way to God Himself. To many traditionalists, the end of this world picture seemed to represent a threat to religion and all cultural values. The English poet John Donne

(1572–1631) expressed it most famously in his poem 'An Anatomie of the World' (1611):

And new Philosophy calls all in doubt,
The Element of fire is quite put out;
The Sun is lost, and th'earth, and no man's wit
Can well direct him where to look for it.

But it wasn't just astronomy that had gone awry...
''Tis all in pieces', he went on, 'all cohaerence gone.'
Similarly, for George Chapman (1559–1634), who wrote in his poem 'The Tears of Peace' (1609):

Heaven moves so far off that men say it stands;
And Earth is turned the true and moving Heaven;
And so 'tis left; and so is all truth driven
From her false bosom; all is left alone,
Till all be ordered with confusion.

If all the old values were left shattered and confused, to a large extent it was the new natural philosophers, for better or worse, who took on the role of repairing or replacing them. First of all, though, they had to show that Copernicus was right.

One physics or two?
Aristotle's supremacy was already under siege when Copernicus published his rival cosmology in 1543, but the following decades saw a number of

astronomical phenomena that seriously weakened Aristotelian cosmology. In a wonderful irony of history an extremely rare astronomical phenomenon happened to occur in 1572. This was the appearance in the sky of what we now call a supernova – a bright burst of light caused by the dying throes of a star. This star had evidently been invisible to the naked eye but as it exploded it became temporarily visible on Earth even in daylight. It immediately attracted the scrutiny of the Danish astronomer Tycho Brahe (later to work with Kepler), who failed to detect any parallax, thereby establishing that it must be, as he thought, a new star that appeared from nowhere. Suddenly, the Ancient Greek assumption that the heavens are perfect and unchanging was shown to be wrong.

Inspired by this bit of debunking, Tycho decided to measure the parallax of a comet that appeared just a few years later. Comets, like meteors, had always been assumed to be meteorological phenomena, so nobody had measured their parallax before. Tycho was able to conclude that the comet was well beyond the sphere of the Moon, and that it too showed the heavens to be changeable. Further observations revealed that the comet must be cutting through the heavenly spheres.

Tycho seems to have believed that Aristotelian cosmologists, and even Copernicus, were committed to a belief in rigid, crystalline spheres, and that his proof of the path of the comet of 1577 entirely shattered this idea. Ironically, the latest research by historians suggests that there was no such commitment but that, after Tycho, the hardness or fluidity of the spheres became a hot topic of debate. The result was an increasing tendency to see the planets as bodies moving freely in space. Even if the notion of spheres was maintained, they were seen as fluid entities in which the planets were distinct entities moving freely, 'like birds in the air, or fish in the sea.' This meant that, instead of a world picture of all-encompassing spheres into which Copernicans had to introduce a small ball revolving independently around the centre, namely the Earth, it was now easier to envisage a space filled with a fluid medium through which various balls, the Earth and the other planets were moving circularly about the centre – the kind of picture that Copernicanism required.

Tycho's published accounts of the new star and the celestial, not atmospheric, nature of the comet of 1577 (which was subsequently confirmed by observations of later comets), and the debate they stimulated, seriously weakened the Aristotelian dichotomy between the perfect, unchanging heavens and a corrupt and changeable Earth.

Since the Sun and the fixed stars were stationary, according to Copernicus, it now became possible to regard the Sun as being just another fixed star.

Giordano Bruno, Dominican priest turned heretic, went further and imagined that there might be other Earth-like planets rotating around other stars, and that these might be inhabited, just like our Earth. This was an idea that scarcely could have occurred to anyone earlier. Nobody would ever have speculated about life on one of the Aristotelian heavenly spheres, unless it was angelic life.

Bruno was burned at the stake for heresy in 1600, but the more 'science-fiction' aspects of his thought were to be dramatically supported in 1610 when Galileo turned the newly invented telescope to the heavens. Even with the poor quality of Galileo's lenses, most who looked through his telescope at the Moon agreed with him that what they saw looked like mountains, landmasses and seas on the Moon. This wasn't a sphere of unchanging quintessence; it was a smaller version of Earth, where things must be made of earth, water, fire and air.

This immediately stimulated a number of literary pieces, the earliest examples of science fiction, in which human life on the Moon was described. But more importantly, it revealed to everyone (Galileo was the equivalent of a best-selling author and his discoveries had a huge impact all over Europe) that a huge lump of rock and water, just like the Earth (albeit a bit smaller), could be flying around through space without coming to a halt, and without coming crashing down to Earth, or to the centre of the universe. If the Moon could do it, and nobody ever denied the motion of the Moon, why shouldn't the Earth be able to do it too?

How the Earth moves

However, it's not enough to suggest that the Earth may be no different from the other heavenly bodies and moving like them. To win the day, the Copernicans really had to show how the Earth could be moving, to explain what causes it to move and to keep on moving incessantly.

This wasn't going to be easy. To do it, the Copernicans had to develop a new theory of dynamics and a new theory of motion. But this is why Copernicus's revolution in astronomy led to a complete revolution in physical science, and ultimately in the scientific worldview. The revolution of the Earth demanded a revolution in science.

One of the first steps was taken in an unexpected place. A physician to Queen Elizabeth I of England, a man by the name of William Gilbert (1540–1603), published the first major study of magnetism in a book called *On the Magnet* (1600). Now, some of Gilbert's best friends were the élite mariners and navigators who were important characters in Elizabeth's royal court in those days – the days of circumnavigations of the globe, of discoveries of new lands and new trade routes, and of the settling of new colonies. Through them, he came to hear of a new discovery made by a retired sailor turned compass-maker, called Robert Norman. Norman discovered what he called magnetic 'dip'.

While making an especially large compass one day, Norman started off, as usual, by ensuring the iron compass needle was perfectly balanced. He then magnetised the needle by rubbing it with a lodestone, a piece of magnetic ore. After this process the needle no longer pivoted on its mounting in a horizontal position but dipped down towards the north. So, the needle doesn't just point to the north, it points down to the ground on its northern side.

Norman was evidently used to this and dealt with it by cutting a bit off the north part of the magnet to restore the balance. On this occasion, he cut too much off and ruined the whole needle. This prompted him to ask the educated mariners of his acquaintance, Gilbert's friends, if they knew what caused this dipping of the magnetised iron. They didn't, but Gilbert was brilliant enough to guess the answer.

At this time it was always assumed that magnets pointed to the pole star, or to the nearby polar axis about which the sky rotated. Hearing about dip, Gilbert immediately guessed that the magnet was pointing to the magnetic pole of the Earth itself. The needle didn't point up to the sky, it pointed down into the Earth. It didn't point north along the surface of the Earth – that is, towards the horizon in a northerly direction – it pointed down below the horizon, straight towards the Earth's pole through the curvature of the Earth. Straight away, therefore, Gilbert concluded that the Earth itself must be a gigantic magnet.

But why did Gilbert think of this, while his clever mariner friends did not? The answer is simple: he was the only one who believed the Copernican theory must be true and who had been pondering night and day how to advance the Copernican cause by showing how the Earth could move. News of the mysterious phenomenon of magnetic dip immediately provided him with the ghost of an answer. Magnets are heavy lumps of earthy material that can quite spontaneously move themselves around, rotating upon a circle.

The whole point of Gilbert's *On the Magnet* is to try to establish that the Earth really is a giant magnet, and to use that to argue that the Earth, just like any other magnet, has the power to turn itself around in a circle. Gilbert was hoping, therefore, that he could at least establish the daily rotation of the Earth on its axis by direct analogy with magnets.

When it came to the Earth's annual revolution about the Sun, Gilbert had to be a bit more speculative. Drawing on the traditional Aristotelian belief that something that can spontaneously move itself must be animated, he insisted that the Earth must have a soul. The Earth's magnetism endows it with self-movement, and so indicates it has a soul. If it has a soul, it can move itself in other ways, besides the rotation on its axis. Gilbert suggested that the power of the Sun, also animated, somehow influenced the Earth and stimulated it to move itself around the sun, along with the other planets.

Over in continental Europe, Gilbert's ideas attracted widespread attention. Two of his biggest fans were Galileo and Kepler, both of whom were looking for a way to explain the motion of the Earth. In the end, Gilbert's ideas proved too magical for the down-to-earth Galileo, and he chose a different path. For Kepler, however, Pythagorean mystic that he was, cosmic magnetism looked just the ticket.

At this point in his career, Kepler was in the laboured process of working out the correct elliptical path for Mars. We've already seen that he explained *why* the planets moved on elliptical orbits at varying speeds by referring to the Pythagorean music of the spheres. But he also needed to explain *how* they could do this. This is where magnetism came in.

All magnets have two poles, north and south, and the Earth is no exception. But suppose the Sun is an exception and only has one pole – say, a north pole. Kepler postulated that the Sun must, like the Earth, be rotating about its own axis, and that, accordingly, lines or spokes of magnetic force emanating from the Sun will continuously sweep around the sky. These lines of force swept the planets around in their orbits.

What's more, if the Earth is in orbit around the Sun with its south pole orientated closer to the Sun than its north pole, then there will be an attractive force operating between the Earth and Sun. The Earth will accelerate as it approaches the Sun. But suppose the orientation of the Earth then shifts in such a way that its north pole swings round closer to the sun than its south pole. A repulsive force will now operate between the Earth and Sun (like poles of a magnet repel, opposites attract). The Earth will tend to move away from the Sun and its moving away will counteract its previous acceleration, resulting in a slowing down of its orbital motion. By the time the Earth gets to its aphelion (furthest point from the Sun) its orientation will have changed again and it will begin to be attracted once more in towards the Sun.

Unable to prove this mathematically, Kepler assumed that the resulting orbit in these circumstances would be an ellipse, with the planet moving faster at the part of its orbit closest to the Sun, and slower when further. This provided Kepler with the explanation of planetary motions he needed for his *New Astronomy* of 1609 (see Figure 12).

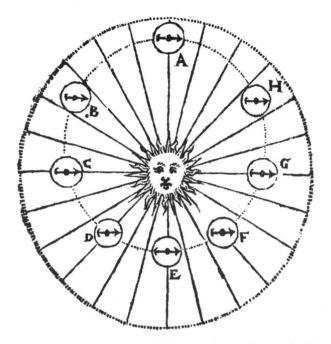

Figure 12. Kepler's attempt to explain elliptical orbits in terms of magnetic attraction and repulsion. The planet is shown in successive positions around the Sun with a superimposed compass needle showing the direction of the planet's magnetic field. Kepler assumed that the Sun acted like a magnet with one pole, which would sometimes attract the planet and sometimes repel it, depending on the position of the planet. At F the planet is still being attracted; at D it is being repelled.

This tradition of what was often called 'magnetical philosophy' proved especially attractive in England, where Gilbert was a local hero. It became influential especially at Oxford University from the 1640s where it was eventually picked up by Christopher Wren (1632–1723), more famous for designing St. Paul's Cathedral, and Robert Hooke (1635–1703), more famous for Hooke's Law.

Contemplating over a number of years about Kepler's attempt to explain planetary motions, using magnetism, Wren and Hooke developed a powerful alternative to the standard accounts. It was usual to try to explain the orbital motions in terms of a centrifugal force, tending to throw moving planets away from the Sun, and a centripetal force, tending to push them into the Sun. The mathematical trick that eluded everyone was to show how these forces balanced to keep the planets in their precise orbits.

Wren and Hooke believed that Kepler's Laws of Planetary Motion could all be explained in terms of a single attractive force towards the Sun, and an assumed straight-line motion of the planet past the Sun. Unlike Kepler, they realised that if the force operating between the Sun and planet was inversely proportional to the square of the distance between them (that is to say, the force diminished as the distance got larger), a passing straight-line motion would be bent round by the attractive force to capture the planet in an elliptical orbit.

Although they sometimes spoke of this attraction as magnetic, Wren and Hooke both knew that the Earth's gravity was not exactly the same as the much more specific magnetic force (which was mostly confined to a few magnetic substances, like iron), but they saw themselves as extending Gilbert's claims. The important point was that the Earth, like magnets, had an 'orb of virtue' or 'sphere of activity' around it, which attracted other bodies.

Unfortunately for Wren and Hooke, neither was a sufficiently expert mathematician to provide a proof of their 'inverse square law'. The necessary proof was provided by Isaac Newton (1642–1727) in his *Mathematical Principles of Natural Philosophy* of 1687, but only after he'd been told of the idea by Wren and Hooke. Newton's book is still acknowledged to be one of the most influential and impressive text books of mathematical physics of all time, and the most important aspect of the book, from which all its other innovations flowed, was its proof of Kepler's laws of planetary motion and the universal principle of gravitation based on the inverse square law. In a very real sense, this all grew out of William Gilbert's attempts to show how the Earth could be moving.

Wren and Hooke managed to dispense with Kepler's awkward notion of spokes of magnetic force from the Sun sweeping the planets around by assuming a passing straight-line motion of a planet that was simply bent into an ellipse by the gravitational force. But where did this idea of passing motion come from? Essentially it came from the alternative attempt to explain how the Earth moved, which had been developed initially by Galileo. Galileo obviously admired Gilbert's ingenuity, but he was too sceptical of magic to ever believe in what is called 'action at a distance', a force capable of operating mysteriously over millions of miles of space. So, if he wanted to explain the motion of the Earth, he had to come up with something else.

What Galileo did was to deliberately remove from his account any suggestion of force, which to him always smacked of magic. Stripping his argument to the bare bones, what he did was to ask his readers to envisage what happens to a perfectly spherical, smoothly polished sphere placed on a glassy smooth slope. Obviously the sphere will begin to roll down the slope gathering speed as it goes. If the sphere is given a sharp push up the slope, by contrast, it will gradually slow down as it moves upwards.

Clearly, force is involved in these cases (the force of gravity). But now imagine if the glassy surface is laid perfectly flat. Now what happens if the sphere is given a sharp push? If we assume that the ball and the surface are so smooth that there is no friction between them, the ball will continue to roll for a long time. In fact, if there really wasn't any friction and no wind resistance, and nothing else to interfere with the movement of the ball, it will carry on rolling at exactly the same speed until it gets to the end of the surface. If the surface could go on forever, so would the motion. There'd be nothing to make it speed up, and nothing to make it slow down, much less to stop. This is effectively a situation where no forces are involved (which is what Galileo wanted). Now, imagine you needed to carry out this experiment for real. How would you make such a perfectly flat surface that continued for mile after mile without ever sloping up or down? If you think about it, your flat surface would have to follow the curvature of the Earth. What we mean by a slope is something that gets further from the centre of the Earth in one direction and closer to it in the other. Something flat, on the Earth, is in fact something that matches the spherical curvature of the Earth. So, if you set a sphere rolling around a frictionless circular path, it will carry on for ever. What's more, it will do so precisely because there are no forces involved, nothing to slow the ball down or to speed it up. We now need just one more step in our argument. God set the ball of the Earth rolling around a circular path when He created the universe, and it's been going around ever since and will always do so, QED.

This is a wonderful argument – especially the way Galileo presents it in his literary masterpiece, *Dialogue on the Two Chief World Systems*, the book that got him into big trouble with the Catholic

Church – but it simply won't work. It would be fine if the planets did move in perfect circles with uniform speeds, but as we know (thanks especially to Kepler), they don't.

Poor old Galileo was so anxious to develop an account of the Earth's motion that avoided the magical notion of force that he was obliged to proceed as if the Platonic dream of uniform circular motions could still be a reality. He had precisely the same difficulties with the tides. The standard belief was that the Moon exerted an occult influence that caused the periodic rising and falling of the seas, and Kepler had recently strongly endorsed this view. For Galileo this magical view had to be dismissed as foolish. Instead, he developed an elaborate explanation of the tides in terms of the sea waters continually slewing about as a result of the Earth's motions. He was very proud of this idea and it played a leading role in his efforts to prove the motions of the Earth. But it was entirely misconceived. As we now know without any doubt whatsoever, the tides are caused by the Moon (and the Sun) exerting an occult influence, which we call gravity. This was finally proven by Isaac Newton, undoubtedly the greatest occult philosopher of all time.

In spite of being wrong, Galileo's attempt to explain the motion of the Earth was to prove extremely fruitful. His argument suggested for the first time that motion might be the sort of thing that will continue indefinitely if left undisturbed. In other words, motion might be just like being at rest. A stationary body will stay put until something moves it. Maybe a moving body will continue moving until something stops it. Another of Aristotle's common-sense beliefs was put to the test. His claim that something moving must be moved every inch of the way by something else now seemed less persuasive. Galileo's theory of perpetual circular motion soon came to be replaced by what is called the 'principle of inertia'. According to this principle, it is motion in a straight line that continues indefinitely, if left unperturbed.

These speculations enabled Wren and Hooke to take as the starting point for their account of planetary motions an assumed straight-line motion of the planets. They didn't need to explain planetary orbits in terms of the balance of two forces, one into the Sun and one away from it. They only had to assume one inward force, analogous to Gilbert's magnetism, and the inertial tendency of the planet to try to keep going in a straight line. The result, as they guessed and as Newton mathematically proved, would be an elliptical orbit in which the speed of the planet alternately accelerated and decelerated.

During the course of his attempts to understand how, contrary to common sense, the motion of the Earth might be perfectly feasible, Galileo completely recast the science of motion. As well as pointing the way to the principle of inertia, he discovered that all bodies fall at the same rate (providing that air resistance does not significantly affect the fall, the way it does in the case of a leaf or a feather). He also established that it is possible for a body to have two movements at once. This was important because the standard Aristotelian account asserted that a body could only move one way at a time, with unnatural or natural motion but not both.

This was another reason for the impossibility of the Copernican theory. Copernicus had to suppose, for example, that an arrow fired straight up into the air not only moved upwards (and subsequently downwards) in a straight line, but also all the while it moved around the centre of the Earth in an easterly direction, moving together with the Earth itself. If this didn't happen, Aristotle would be right and the arrow fired straight up would eventually land far to the west of where it was fired. Galileo was able to show that the motion of a body could be continuously compounded of a 'forced' motion (caused, in this instance, by firing a bow) and a 'natural' motion caused by the rotation of the Earth.

Space: the final frontier
Galileo made another important contribution to the story. We've already seen that he used the telescope to reveal what he thought were landmasses and seas on the Moon, so bringing to an end the

idea that things in the heavens were categorically different from the Earth. Keeping the telescope to his eye, he also noticed there were far more stars in the sky than you could see with the naked eye. Straight away, this suggested that some stars were much further away than others. But there was something else…

Through the telescope, the stars, even the ones visible to the naked eye, looked just the same; they continued to look like twinkling pricks of light. Everything else in the telescope was magnified. The planets looked distinctly disc-like and larger than with the naked eye, but the stars continued to look like mere points of light. What could this mean?

Galileo used this puzzling result to confirm the Copernican insistence that the stars must be many times further away from the Earth than the Ptolemaic system required them to be. The lack of discernible parallax in observations of the stars, even though the Earth was supposed to move many millions of miles through space, only made sense if the stars were inconceivably far away. To all but the few who excitedly embraced an infinite universe, this had always seemed ridiculously implausible. But now the telescope seemed to confirm that it was true after all. Natural philosophers looked again at the arguments of those who said the stars must be scattered throughout an infinite space and began to find them convincing.

But this infinite space raised another problem. In the Aristotelian universe the heavenly spheres were all contiguous, touching one another like the layers of an onion, with no gaps in between. There couldn't be any gaps because Aristotle believed that a vacuum was a physical impossibility. For Aristotle the concept of empty space made no sense. Space was defined by the body or bodies that occupied it, if there were no bodies, it made no sense to talk about the space where those bodies weren't; no bodies, no space. But contiguous spheres presented no problems for Ptolemaic astronomy. Ptolemy could simply assume that the sphere of the fixed stars was enveloped contiguously around the outside of the sphere of Saturn. But if the sphere of fixed stars had to be vastly far removed from the sphere of Saturn, what lay in between them? Besides, the realisations that comets were astronomical phenomena, and that the stars were not fixed on a sphere but scattered at different distances from us, were already tending to suggest that there were no spheres anyway. So, what was out there? Maybe all that infinite space was mostly empty after all.

The possibility, or not, of the existence of a vacuum became a growth area in natural philosophy after Copernicus. The first air-pump, for creating artificial vacua, was invented in 1650 by the Burgomaster of Magdeburg in Germany,

Ottovon Guericke (1602–86). Experiments with air-pumps soon took on a life of their own and led to major reforms of scientific knowledge in a number of areas, but the fact remains that this experimental research grew out of the Copernican revolution. Guericke makes it perfectly clear in his book, *New Experiments on Empty Space*, that his main concern in beginning his experiments was to provide support for the Copernican theory.

Remarkably, Guericke also invented the first generator of static electricity (a ball of fused sulphur that developed a static charge when rubbed) to provide further evidence in favour of Copernicus. Guericke shows how his sulphur globe can attract various small objects, which will then remain on its surface as the globe is spun round. 'Now we can see', Guericke wrote, 'how the sphere of our Earth holds and maintains all animals and other bodies on its surface and carries them about with it in its daily 24-hour motion.'

The electrical phenomena Guericke described attracted immediate attention and, again, led to major new understandings of the natural world, but these new developments also arose initially from Copernicanism.

The importance of the Copernican stimulus to a new understanding of space – as infinite and empty – can be seen in Isaac Newton's concept of 'absolute space'. Newton's laws of motion are only valid on the assumption that there is a real, unchanging, non-interacting, infinite empty space that provides the arena in which all bodies move and interact. Unlike Aristotelian space, in which there are so-called natural places for each of the five elements, and in which natural movements are completely different in different regions (straight-line natural motions below the Moon, circular natural motions above it), Newton's space has to be everywhere the same throughout its infinite expanse, undifferentiated, unaffected by the bodies in it, and not affecting them – in short, an absolute space.

When Newton wrote his *Mathematical Principles of Natural Philosophy* he began by defining the technical terms and concepts he was about to use. In spite of the crucial importance of the notion of absolute space to his physics, however, he did not have to define it. 'I do not define absolute space', he simply wrote, 'as being well known to all.' By 1687 it was indeed well known, but only as a result of Copernicus and the demands his system made for a reform of previous views of space.

• CHAPTER 6 •

LAST WORDS

So, Copernican astronomy led to a completely new physics. Attempts to explain the motion of the Earth led, through Gilbert's and Kepler's ideas on magnetism, to a theory of universal gravitational attraction, and, through Galileo's perpetual circular motion, to a theory of inertia. These ideas, as well as stimulating further developments in other areas, could then be combined in the absolute space of the Copernican universe to constitute the first really successful system of mathematical physics, as codified by Isaac Newton, and which was to form the basis of modern physical science.

However, its influence was even more far-reaching. Establishment of the Copernican theory also helped to separate science from common sense by showing that the physical world didn't have to be just the way it seemed – it could be far more interesting than that. Just as the Earth didn't have to be stationary, so natural philosophers soon began to suggest that the bodies we perceive, with colours, smells and tastes, might be made out of imperceptible atoms that have no colours, or tastes or smells, but simply give rise to these impressions in us by the way they combine with one another. This, in turn, could lead to the belief that maybe life itself (and even one's mind) might be produced by the way atoms combine with one another.

Going hand in hand with the move away from common sense was the rise of mathematics as a way of establishing how the world works. Previously, natural philosophers boasted on the one hand that they could explain things in terms of physical causes, giving reasons why things occurred the way they did, and dismissed mathematics on the other hand because it couldn't explain anything; it could only provide its own kind of technical description.

After Copernicus, natural philosophers increasingly recognised the importance of mathematics for truly explaining the world. We saw earlier that Christopher Wren and Robert Hooke arrived at the inverse square law of gravitational attraction and suggested it could explain Kepler's three laws of planetary motion, then told Newton about it. But it is Newton who has always been credited with the idea, not Wren or Hooke. Why? Because it was Newton who showed that it was true, and that's what wins the Brownie points. Hooke did actually demand some acknowledgement in Newton's book that the principle was his, not Newton's, but to no avail. Newton was outraged at the suggestion that a great mathematician like himself should acknowledge the importance of someone who could go no further with a lucky guess.

A hundred years earlier, Hooke might have been seen as the great natural philosopher who discovered the universal principle of gravitation, while a less well-remembered Newton might have been recorded as the mindless calculator who had the knack of doing the sums to show that Hooke was right. This is how it had been when mathematicians were regarded as greatly inferior to natural philosophers. But by the late 17th century, things had changed. Hooke's contemporaries, like us, were in awe of mathematicians, and it was what mathematicians did that counted. Admittedly, Newton was a pretty awesome character in his own right, but the general attitude to the value of mathematics and mathematicians had begun to emerge many years before, as a result of the astronomy of Nicolaus Copernicus.

No other single contribution to the history of science was to have the far-reaching consequences that followed from Copernicus's decision to set the Earth revolving with the other planets about the Sun. These consequences went far beyond science. Sigmund Freud (1856–1939), founder of psychoanalysis, popularised the idea that Copernicus had caused consternation by moving mankind from its privileged position at the centre of the universe, but there is little sense of that in the historical record.

Certainly a few poets and other intellectuals were confused and dismayed by the possible implications of Copernican theory for traditional values, but there were many more who were clearly excited by the prospect of such a different world.

It used to be thought that Copernicanism spelled the end of astrology and inaugurated the disenchantment of the old magical worldview and the rise of rational science. But astrologers demanding a more reliable astronomy were among the first to embrace Copernican reforms, and Hermetic magicians liked the way Copernicus replaced man at the centre of the world system by the Sun, visible analogue of God.

Magical belief in the music of the spheres played a major role in confirming Kepler and Newton in their Copernicanism. Furthermore, for them and for others, the mathematical harmony of the Copernican cosmos showed that it must have been created by a divine intelligence. For many, however, the Copernican theory showed that the Bible and the doctrines of the Church could be based on misconceptions, and moving mankind from the centre of the universe was for them a triumphant confirmation that there was no meaning in any position in a Godless universe.

Freud should have known better; it takes all sorts to make a world. And if anything is certain, apart from death and taxes, it's the fact that Copernicus helped to make our world.

GLOSSARY

Aether: frequently used term to denote the substance of the heavens, otherwise known as the fifth element, or quintessence.

Apogee (see also **Perigee):** the furthest point on a planet's elliptical orbit from the Sun, which is also the point at which its orbital speed is at a minimum.

Aristotelian: the philosophy of Aristotle (384–322 BC), whose works were most well known and so most influential in Western Europe in the Middle Ages. Used loosely here to include even medieval versions of his philosophy, which varied slightly from his original.

Art: system of craft skills used for practical purposes. Distinguished from science, which was concerned with reliable knowledge, because arts were known to work or be effective even though nobody knew why. Some aspects of human practice exhibited features of both science and art. Medicine was sometimes based on a body of knowledge and theory, although often healing was achieved mysteriously by craft know-how, but without knowing why. Similarly, astronomy displayed elements of science and art.

Astrology: interpretative art of judging how the configuration of the heavens might affect conditions on Earth. Widely believed to be valid from Ancient times through to the 17th century, even though interpretations were acknowledged often to be mistaken.

Astronomy: science or art of calculating movements of heavenly bodies based on successive positions determined by observation.

Atomism: philosophy first developed among the Ancient Greeks and revived in the Renaissance that explained all physical phenomena in terms of the movements, collisions, interactions and combinations of particles of matter that were so small they were held to be indivisible.

Axis: the line through the diameter of a rotating sphere about which it is said to rotate. The axis remains fixed, as the sphere turns around it.

Big Bang: a theoretically supposed creation event in which all the matter and radiation in the universe are believed to have come into existence in an explosion at a finite time in the past (between 10 thousand and 20 thousand million years ago). The theory has been very successful in explaining a number of cosmic phenomena, including the expansion of the universe.

Calendar: system of assigning regularities to the passage of the seasons and the years, generally based on trying to reconcile the cycles of the Sun with the cycles of the Moon.

Canon: an administrative position in the Church; a cathedral clerk, closer to a civil servant than to a priest.

Celestial equator: equator of the theorised 'celestial sphere'. A great circle dividing the sphere into two equal hemispheres, north and south, just as the Earth's equator divides it into two halves.

Conjunction: situation in the heavens when two planets lie in the same direction from the Earth – that is, they look as if they are together against the backdrop of the fixed stars.

Constellation: group of stars considered by early astronomers to lie within the outline of an animal, object or mythological character, imagined on the heavenly vault. Used as a means of remembering stellar positions, and to describe heavenly locations, as well as being endowed with astrological significance.

Copernicanism: belief in the cosmology of Nicolaus Copernicus (1473–1543) in which the Earth, along with the other planets, revolves around the Sun and rotates on its axis.

Cosmology: science of understanding the presupposed harmonious and unified structure and arrangement of the heavens.

Deferent (see also **Epicycle):** the great circle carrying with a uniform rotational speed the centre of a planet's epicycle, which is itself centred on or near the Earth in the Ptolemaic system, or on or near the Sun in the Copernican system.

Eccentric: great circle carrying a planet or a planet's epicycle with uniform speed but whose centre is displaced from the Earth itself in the Ptolemaic system, or from the Sun in the Copernican system – in other words, an off-centre circle.

Element: Aristotle believed that all earthly bodies were composed of combinations of the four elements (earth, water, air and fire), and that the heavens were composed of a fifth element, never found on Earth, variously known as the quintessence, or aether.

Ellipse: a closed curve, like a stretched-out circle. A circle is a curved line around a single point that always maintains the same distance away from that point. An ellipse is a curved line around two points (the foci) which maintains a constant relationship to both points such that the distances from any point on the curve to each of the two foci always add up to the same distance. It's the kind of curve you see when you look at a circular shape, say the top of a cup or glass, from an angle (see Figure 9).

Ephemerides: published tables of solar, lunar and planetary positions as seen from a particular position over a designated period (usually annual), based almost entirely on calculation (perhaps checked by the odd observation here and there), and intended to allow astrological readings and navigation without having to make one's own astronomical observations.

Epicycle (see also **Deferent):** secondary circle around which a heavenly body moves with uniform speed, which is itself centred on a larger circle – the deferent – that is turning with uniform speed around a point located at or near the Earth in the Ptolemaic system, and at or near the Sun in the Copernican system – in other words, a circle on a circle.

Equant: a point in space on the opposite side of the Earth from the centre of rotation of a planet's eccentric, from which it is supposed that the motion of the planet would look uniform in speed. Used by Claudius Ptolemy (*c.* AD 100–170) to help take account of the fact that planets speed up and slow down in their orbits.

Equinox: the point at which the sun crosses the celestial equator during its apparent annual motion. It crosses from south to north of the equator at the so-called vernal equinox (in summer in the northern hemisphere), and from north to south at the autumnal equinox.

Eudoxan: relating to the astronomical system of Eudoxus of Cnidus (*c.* 400–347 BC), in which all heavenly movements were described in terms of the interacting motions of nested spheres all revolving around the same centre but on differing axes of rotation, with differing uniform speeds and differing directions.

Fixed stars: points of light visible in the night sky called 'fixed' to distinguish them from the 'wandering stars', or planets. The fixed stars were believed to move around the Earth once every day, but to remain fixed in their positions with regard to one another. Copernicus made them even more fixed by attributing their daily movement to the rotation of the Earth.

Geometrical archetype: according to Kepler, the blueprint used by God to determine where to place the planets and how many planets to place. Consisted of alternately nesting one of the five Platonic solids between each of the planetary spheres.

Hippopede: name given by Eudoxus to the loop made by a planet in retrograde motion according to his scheme using the combined motions of homocentric spheres. Named after a hobble used on horses' feet.

Holy Trinity: one of the central beliefs of Christianity, although denied by some Christian sects, that the divine being is mysteriously three beings in one: Father, Son and Holy Spirit. It derives from early attempts to understand the relationship of the divinity of Jesus Christ to the divinity of God without rejecting monotheistic assumptions that there is only one God.

Hooke's Law: law of elasticity discovered by Robert Hooke (1635–1703) in 1660, which states that extension (or other deformation of shape) of an elastic body is directly proportional to the force applied.

Inertia: property of a body that enables it to resist changes to its motion or to its state of rest. Aristotle erroneously believed that everything that moves must be continuously kept in motion by a mover, but 17th-century thinkers began to realise that just as a body at rest stays put, so a body in motion continues to move uniformly until something stops it or alters its motion.

Kepler's Laws of Planetary Motion: Johannes Kepler (1571–1630) discovered that planets move on elliptical orbits, that the line joining a planet to the Sun sweeps out equal areas in equal periods of time (see Figure 9), and that there is a constant relationship between the sidereal period of a planet and its average distance from the Sun. These laws remain valid today, and enabled a full understanding of planetary astronomy and much more besides.

Liberal arts: seven preliminary subjects taught in medieval and Renaissance European schools and universities. Divided into the *trivium* (grammar, rhetoric and logic) and the *quadrivium* (arithmetic, geometry, astronomy and music).

Model (as in geometrical model): any attempt to envisage the means by which partial or sporadic observations could relate to physical realities. So, the retrograde motion of a planet could be modelled by assuming it was located on an epicycle moving around a larger circle centred on the Earth.

Musical archetype: according to Kepler, the musical scheme used by God to establish the positions and motions of the planets around the Sun, which He used in conjunction with the geometrical archetype in creating the universe.

Natural philosophy: theories and beliefs about the nature of the physical universe. Strictly speaking there was no such thing as 'science' in our modern sense during the period discussed in this book. The term 'natural philosophy' was the closest thing to it, but was concerned with understanding the world rather than trying to change or exploit it. Throughout most of the period covered in this book, Aristotelian versions of natural philosophy were entirely dominant.

Perigee (see also **Apogee):** point on a planet's elliptical orbit where it is closest to the Sun, and where its rotational speed is at a maximum.

Planet: a wandering star, distinguishable by its independent movements across the backdrop of the fixed stars. In this book we are concerned only with the planets visible to the naked eye, which were known since Ancient times – Mercury, Venus, Mars, Jupiter and Saturn.

Platonic: relating to the philosophy of Plato (*c*. 428–347 BC), one of the most influential of the Ancient Greek philosophers.

Platonism: the philosophy of Plato, used here loosely to include later thinkers whose philosophy was derived from Plato but was often so modified that it is now more usually distinguished from Platonism by the label Neoplatonism. This distinction is unnecessary here.

Precession of the equinoxes: slow movement

westwards of the points where the Sun crosses the celestial equator each year. Assumed since Hipparchus (*c.* 190–120 BC), who discovered it, to be a motion of the heavens, but explained by Copernicus by assuming another motion of the Earth – the slow movement of its axis of rotation.

Ptolemaic: relating to the Earth-centred astronomy of Ptolemy, the most influential astronomer before Copernicus.

Pythagorean: relating to the ideas and beliefs of Pythagoras (6th century BC) and his followers. Renowned as a magician and mathematician who believed that the world could be understood in mathematical terms, Pythagoras was also believed to have subscribed to a moving Earth. Accordingly, the Copernican theory was sometimes referred to as the Pythagorean theory.

Quintessence: see entries for **Aether** and **Element**.

Renaissance: period from roughly the 15th century to the end of the 16th century, which saw huge changes in the cultural and economic life of Western Europe, major technological innovations (especially printing, the magnetic compass and gunpowder), and an extensive revival of the learning and literature of Ancient Greece and Rome.

Retrograde motion: period during which a planet seems to be moving backwards in the sky, rather than in its usual direction. Since Copernicus, it's now known to be an illusion caused by the Earth overtaking the planet, but was explained by Ptolemy as the result of the planet looping-the-loop on its epicycle (see Figures 3 and 4).

Revolution: used here consistently to refer to the movement of a planet around the Earth (in the Ptolemaic system) or around the Sun (in the Copernican system).

Rotation: used here consistently to refer to the movement of a sphere, or the Earth, around its own axis.

Scepticism: Ancient philosophy revived during the Renaissance and gaining new life of its own, in which all rational arguments are regarded as subjective, contentious and unreliable.

Science: often used here in the modern sense but strictly speaking it is anachronistic during the period covered in this book since there was no such thing as modern science then (see **Natural philosophy**). In this period the term 'science' equated with certain or reliable knowledge – so, the 'science of theology' would be as natural a usage as the 'science of optics' or the 'mathematical sciences'. See also **Art**.

Sidereal period: time taken by a planet to complete one revolution around its centre of rotation, measured by reference to the background of the fixed stars (that is, by determining the time taken for the planet to return to a particular star in the sky).

Sphere: a great sphere in the heavens completely surrounding the central body – whether that be the Earth, or the Sun and all other heavenly spheres below it – derived ultimately from the Eudoxan cosmology, but adopted by Aristotle and remaining influential beyond Copernicus. The sphere of Mars, therefore, does not refer to the red planet that so fascinates us, but to the vast sphere enveloping everything in the universe below it, and on which the planet itself is nothing more than a reflective spot.

Stoicism: the philosophy of the Stoics, a group of Ancient Greek thinkers in the period after Aristotle, who developed a characteristic philosophy, including teachings on morality and the nature of the physical universe.

Further Reading

There are a number of general histories of astronomy that can be recommended.

The *Cambridge Illustrated History of Astronomy*, edited by Michael Hoskin (Cambridge: Cambridge University Press, 1997), is excellent, giving a broad coverage from the Ancient Greeks to the Copernican Revolution and beyond. This publication includes a really useful chapter on Islamic astronomy, which had an important influence on medieval European astronomy but which has not been mentioned here. Additionally, it contains much information about the role and nature of instruments in astronomy.

The Fontana History of Astronomy and Cosmology, by John North (London: Fontana Press, 1994), provides an even wider coverage of the history of astronomy in other periods and cultures (but without pictures), and includes excellent detail on Copernicus, his background and his subsequent impact.

For more idiosyncratic but entertaining accounts, see Rocky Kolb's *Blind Watchers of the Sky: The People and Ideas that Shaped Our View of the Universe* (Oxford: Oxford University Press, 1999), and Owen Gingerich's episodic *Great Copernicus Chase and Other Adventures in Astronomical History* (Cambridge: Cambridge University Press, 1992).

If you are not daunted by detail and want to get a bit more technical, the classic treatment is Thomas S. Kuhn, *The Copernican Revolution: Planetary Astronomy in the Development of Western Thought* (Cambridge, MA: Harvard University Press, 1957).

There are excellent chapters on Ancient and medieval astronomy up to and including Copernicus in *Early Physics and Astronomy: A Historical Introduction*, by Olaf Pedersen and Mogens Pihl (Cambridge: Cambridge University Press, 1993). This book also allows the reader to understand how work in astronomy related to contemporary developments in physics and other mathematical sciences.

The General History of Astronomy, Volume 2A: Planetary Astronomy from the Renaissance to the Rise of Astrophysics, edited by René Taton and Curtis Wilson (Cambridge: Cambridge University Press, 1989), is excellent, but takes as its starting point the period immediately after Copernicus. There is a fascinating collection of advanced articles in *The Copernican Achievement*, edited by Robert S. Westman (Berkeley: University of California Press, 1975).

If you are interested in pursuing some of the byways mentioned throughout the course of this book you might want to look at the classic account of the development of the infinite universe by Alexandre Koyré, *From the Closed World to the Infinite Universe* (Baltimore: Johns Hopkins University Press, 1957).

For a more technical account of how astronomers established cosmic dimensions, and much else besides, see Albert Van Helden's *Measuring the Universe: Cosmic Dimensions from Aristarchus to Halley* (Chicago: University of Chicago Press, 1985).

On various attempts to reform the calendar see E.G. Richards, *Mapping Time: The Calendar and Its History* (Oxford: Oxford University Press, 1998).

The role of William Gilbert and his magnetical philosophy is nicely covered in articles by Stephen Pumfrey and James R. Bennett in *The General History of Astronomy, Volume 2A*, mentioned above.

There is an excellent biography of Tycho Brahe – Victor E. Thoren's *Tycho Brahe: The Lord of Uraniborg* (Cambridge: Cambridge University Press, 1990).

For a simple introduction to Kepler, see James R. Voelkel, *Johannes Kepler and the New Astronomy* (Oxford: Oxford University Press, 1999).

For more advanced studies, J.V. Field's *Kepler's Geometrical Cosmology* (London: Athlone Press, 1988) and Bruce Stephenson's *Kepler's Physical Astronomy* (Princeton: Princeton University Press, 1994) cannot be bettered.

There is no shortage of books on Galileo. A good recent introduction is David Wootton, *Galileo, Watcher of the Skies* (New Haven and London: Yale University Press, 2010). For an accessible set of essays on different aspects of Newton's work, including his interest in magic and Pythagorean music of the spheres as well as his more technical accomplishments, see *Let Newton Be!*, edited by J. Fauvel, R. Flood, M. Shortland and R. Wilson (Oxford: Oxford University Press, 1988).

A concise account of the new physics arising out of Copernicus's astronomy is *The Birth of a New Physics*, by I. Bernard Cohen (Harmondsworth: Penguin, 1987). The best surveys of relations between science and religion, including considerations of Copernicus, Galileo, Kepler and Newton, are *God and Nature: Historical Essays on the Encounter between Christianity and Science*, edited by D.C. Lindberg and R.L. Numbers (Berkeley: University of California Press, 1986), and *Science and Religion: Some Historical Perspectives*, by John Hedley Brooke (Cambridge: Cambridge University Press, 1991).

The shift during the Renaissance in the intellectual status of mathematics and the crucial importance of this for understanding Copernicus's achievement was first noticed by Robert S. Westman, but his important publications are mostly confined to academic journals.

For developments and extensions of his ideas, however, consider James M. Lattis's *Between Copernicus and Galileo: Christoph Clavius and the Collapse of Ptolemaic Cosmology* (Chicago: University of Chicago Press, 1994), and Peter Dear's *Discipline and Experience: The Mathematical Way in the Scientific Revolution* (Chicago: University of Chicago Press, 1995).

ICONSCIENCE

THE ICON SCIENCE 25TH ANNIVERSARY SERIES IS A COLLECTION OF BOOKS ON GROUNDBREAKING MOMENTS IN SCIENCE HISTORY, PUBLISHED THROUGHOUT 2017

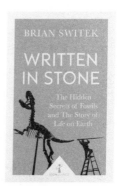

The Comet Sweeper
9781785781667

Eureka!
9781785781919

Written in Stone
9781785782015
(not available in North America)

Science and Islam
9781785782022

Atom
9781785782053

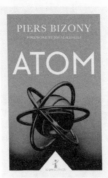

An Entertainment for Angels
9781785782077
(not available in North America)

Sex, Botany and Empire
9781785782275
(not available in North America)

Knowledge is Power
9781785782367

Turing and the Universal Machine
9781785782381

Frank Whittle and the
Invention of the Jet
9781785782411

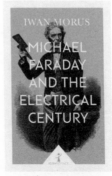

Michael Faraday and the
Electrical Century
9781785782671

Moving Heaven and Earth
9781785782695